Edward Tyson

Orang-Outang, Sive Homo Sylvestris

Or, the anatomy of a pygmie compared with that of a monkey, an ape, and a man.

To which is added, A philological essay concerning the pygmies, the cynocephali,

the satyrs, and sphinges of the ancients

Edward Tyson

Orang-Outang, Sive Homo Sylvestris
Or, the anatomy of a pygmie compared with that of a monkey, an ape, and a man. To which is added, A philological essay concerning the pygmies, the cynocephali, the satyrs, and sphinges of the ancients

ISBN/EAN: 9783337393328

Printed in Europe, USA, Canada, Australia, Japan

Cover: Foto ©berggeist007 / pixelio.de

More available books at **www.hansebooks.com**

Orang-Outang, sive Homo Sylvestris:

OR, THE

ANATOMY

OF A

PYGMIE

Compared with that of a

Monkey, an *Ape,* and a *Man.*

To which is added, A

PHILOLOGICAL ESSAY

Concerning the

Pygmies, the *Cynocephali,* the *Satyrs,* and *Sphinges*
of the ANCIENTS.

Wherein it will appear that they are all either *APES* or
MONKEYS, and not *MEN,* as formerly pretended.

By *EDWARD TYSON* M. D.

Fellow of the Colledge of Physicians, and the Royal Society :
Physician to the Hospital of *Bethlem* , and Reader of
Anatomy at *Chirurgeons-Hall.*

LONDON:

Printed for *Thomas Bennet* at the *Half-Moon* in St. *Paul's* Church-yard ;
and *Daniel Brown* at the *Black Swan* and *Bible* without *Temple-Bar*
and are to be had of Mr. *Hunt* at the Repository in *Gresham-Colledge.*
M DC XCIX.

J O H N Lord Sommers,

Baron of *EVESHAM,*

Lord High Chancellor

O F

ENGLAND,

One of the Lords of his MAJESTIES
moſt Honourable PRIVY COUNCIL,

And Preſident of the ROYAL SOCIETY.

S I R,

THE great *Variety of weighty and important*
Affairs,in which your Lordſhip is engaged;one
would think, did ſo entirely engroſs your Time,
that you could not have a Minute left to beſtow on the
Muſes. Your unwearied and ſuccesful Application to
the

the Buſineſs *of the* State, *in the* niceſt Conjunctions, *that perhaps* England *ever ſaw* ; *as well as your in-expreſſible* Labours *in diſtributing* Juſtice, *in your High Station* ; *have been attended with* Univerſal Applauſe: *and have convinced all the* World, *how much the* Honour *of his* Majeſty's Government, *and the* Happineſs *of his* People, *depend on the Capacity and Integrity of his* Miniſters. *You have not ſuffered, even neceſſary Refreſhments to interrupt your Conſtant Cares for the Pub-lick. To ſerve your Country, you have defrauded your ſelf both of* Meat, *and* Reſt ; *which, my Lord, is the only* Act *of* Injuſtice, *that was ever charged upon you. Your immoderate* Labours *make daily* Encroachments *upon your* Health ; *or at leaſt 'tis the fear of every good* Man, *that they ſhould. And yet your Lordſhip, notwithſtanding all* Diſſwaſions, *perſeveres inflexible* ; *as if, animated by the Noble Spirit of an Old* Roman, *you were reſolved to Sacrifice your* Life, *for the Good of your* Country.

And yet, my Lord, amidſt ſuch a multiplicity of the greateſt Affairs, *to which you pay a conſtant* Attendance ; *you have not only found* Time, *to apply your* Thoughts *to all kinds of* Literature, *ſo as to become a great* Maſter *in all* ; *But you have likewiſe extended your* Care *to the* Intereſts *of* Learning, *and to the* Encouraging *of thoſe, who ſtudy the* Advancement *of it. Among many other* Inſtances, *your Lordſhip has lately* condeſcended, *to* Preſide *over the* Royal Society;

that
USI

that was inſtituted, for the Improvement of Natural Experimental Philoſophy ; and you have taken care, to Expreſs your great Zeal and Readineſs, to contribute every thing in your Power, to Protect their Intereſts, and Promote their Reputation. And under your Lord-ſhip's enlivening Influence, there is all the Reaſon in the World to expect, that Learning will again flouriſh there, as well as among other Orders of Men.

This, my Lord, has ſo embolden'd me, humbly to pre-ſent this Performance to your Lordſhip. For ſince ſo Great a Patron of Letters is riſen in the midſt of us ; we think we have a ſort of Right to his Countenance and Protection. I wiſh the Preſent I preſume to make you, was more worthy of your Lordſhip : All that I can ſay to recommend it, is, that the Subject is Novel, and that Care has been taken to give it a juſt Deſcription ; which, I may ſay, without vanity, never yet appeared in Publick.

'Tis a true Remark, which we cannot make without Admiration; That from Minerals, to Plants; from Plants, to Animals; and from Animals, to Men ; the Tranſition is ſo gradual, that there appears a very great Simili-tude, as well between the meaneſt Plant, and ſome Mi-nerals ; as between the loweſt Rank of Men, and the higheſt kind of Animals. The Animal of which I have given the Anatomy, coming neareſt to Mankind ; ſeems the Nexus of the Animal and Rational , as your Lord-ſhip, and thoſe of your High Rank and Order for Know-ledge

ledge and Wifdom, approaching neareft to that kind of Beings which is next above us ; Conneɛt the Vifible, and Invifible World.

If this Performance fhall Promote the Defign of the Society, of which I have the Honour to be a Member, and which your Lordfhip is pleafed to Prefide over; by improving the Natural Hiftory of Animals, and affording the Reader any Delightful and Ufeful Inftruɛtions ; I fhall look on my Time and Pains, well rewarded. I am

My Lord,

Your Lordſhip's moſt humble

and moſt obedient Servant

EDWARD TYSON.

THE

PREFACE.

LEAST this Difcourfe fhould be rejected meerly for the Title's fake, as if 'twere intended only to divert the Reader, with the Recital of the Fabulous and Romantick Stories, which have been related on the Subjects I have propofed to treat of: I think it neceffary to premife, that as my chief Defign in this Undertaking is the Improvement of the *Natural Hiftory* of *Animals*; fo I have made it my Bufinefs more, to find out the *Truth*, than to enlarge in the *Mythology*; to inform the *Judgment*, than to pleafe the *Phancy*. And the *Orang-Outang* (whofe *Anatomy* I here give) being a Creature fo very remarkable, and rare; and not only in its External Shape, but much more in the Conformation of a great many of the inward *Vifcera*, fo much refembling a Man; I thought I could not be too particular, in my Defcription of it; though to fome, who have not a Taft of thefe Matters, I may feem prolix and tedious.

To render this *Difquifition* more ufeful, I have made a *Comparative* Survey of this *Animal*, with a *Monkey*, an *Ape*, and a *Man*. By viewing the fame Parts of all thefe together, we may the better obferve *Nature's Gradation* in the Formation of *Animal* Bodies, and the Tranfitions made from one to another; than which, nothing can more conduce to the

A   At-

Attainment of the true Knowledge, both of the *Fabrick*, and *Uses* of the Parts. By following *Nature's* Clew in this wonderful *Labyrinth* of the *Creation*, we may be more eafily admitted into her *Secret Receffes*, which Thread if we mifs, we muft needs err and be bewilder'd.

In drawing up this *Comparifon*, I have made ufe of the Anatomy which is given of *Apes* and *Monkeys* by other Authors; and very frequently have quoted their own words, which has render'd my Difcourfe much longer: For not having thefe *Animals* by me to diffeth and compare, I thought it but juft to let the *Reader* fee, upon what Authorities I went. And though a fhort Reference might be efteemed fufficient, without this tedious and unfafhionable way of inferting the whole *Text*; yet if any one will give himfelf the trouble of Examining the Evidences I have produced, I think I have dealt more kindly by him, in making him a Judge himfelf; than in leaving him barely to truft to my Report. For there are none, who have been converfant with Books, but muft acknowledge, that they have been often impofed upon, for want of this fair dealing, as I have my felf Experienced in this prefent Enquiry. To avoid therefore this Error, my Caution it may be has lead me into another, which I hope the *Reader* will pardon, if he judges it fuch.

Galen formerly diffected *Apes* and *Monkeys*, and recommended to his Scholars the frequent *Anatomizing* them, as ufeful for the attaining the Knowledge of the Structure of the Parts in *Humane* Bodies. Had he met with our *Animal*, it had ferved his turn much better: Nor had he been liable to fome Miftakes, which *Vefalius* charges him with, fince in fo many Parts, the *Orang-Outang* imitates a *Man*, more than *Apes* and *Monkeys* do. Not only *Galen*, but the greateft *Anatomifts* we have had in this laft Age, have exercifed their Pens about them; as plainly appears in the enfuing Difcourfe, which fufficiently juftifies me for engaging in this Argument: I wifh I had fo good an Apology for my Performance.

This

This great Agreement, which I obferved between the *Orang-Outang,* and a *Man,* put me upon confidering, whether it might not afford the Occafion to the Ancients, of inventing the many Relations, which they have given us of feveral *forts* of *Men,* which are no where to be met with but in their Writings. For I could not but think; there might be fome Real Foundation for their *Mythology*; which made me more ftrictly enquire into their Records; and examining them, I always found fomething new, that infenfibly lead me on far beyond what at firft I intended: and if I do not deceive my felf, I have at laft gained a clearer Light in thefe Matters, than any that has hitherto appeared.

For what created the greateft difficulty, was their calling them *Men,* but yet with an Epithet for diftinction fake; as the Ἄνδρες Ἄγριοι, Μικροὶ, Πυγμαῖοι, Μέλανες; fo the Ἄνθρωποι Κυνοπρόσωποι, &c. *i. e.* the *Wild Men,* the *Little Men,* the *Pygmæan Men,* the *Black Men,* the *Men* with *Dogs Faces,* &c. yet at the fame time I find that they made them θηρία, *Wild Beafts*; and if fo, no doubt but they were of the *Quadru-manus* kind; *i. e.* either *Apes* or *Monkeys.* And fuch were likewife the *Satyrs,* the *Fauni, Pan, Ægipan, Sylvanus, Silenus,* and the *Nymphæ,* as alfo the *Sphinges* of the Ancients.

But fo many *Romances* have been made about them, that not only *Strabo* formerly, but the moft noted Men of Learning of late, have looked upon them as meer Fictions of the *Poets,* and have utterly denied them any real Being. *Homer's Geranomachia* therefore, or *Fight* of the *Cranes* and *Pygmies,* I have rendered a probable Story. *Ariftotle's* affertion of the being of *Pygmies,* I have vindicated from the falfe Gloffes of others. The Conjectures of other Learned Men about them, I have examined: And by what I have faid in the following *Philological Effay,* I think I have fully proved, that there were fuch *Animals* as the Ancients called *Pygmies, Cynocephali, Satyrs,* and *Sphinges*; and that they were only *Apes* and *Monkeys.*

Had.

Had my Leisure been greater, I had contracted the whole, and taken more care both in the *Method*, and *Expression*. But most of the vacant Hours from the necessary Attendance on the Business of my Profession, being taken up in Collecting Materials; to gratifie the Importunity of my Friends, who constantly urged the Publication, I sent my Papers Sheet by Sheet to the Press, as I had time to transcribe them; so that I had not a view of them together, till they were printed. If I have discovered the Truth, 'twas what I aimed at, which always appears best, when least disguised; and it has been my chief Care in this Undertaking to pull off those Vails and Masks, wherewith the Poets and Poetical Historians have hitherto obscured it.

Orang-

Orang-Outang sive Homo Sylvestris:

OR, THE

ANATOMY

OF A

PYGMIE.

THAT the *Pygmies* of the Antients were a fort of *Apes,* and not of *Humane Race,* I fhall endeavour to prove in the following *Effay.* And if the *Pygmies* were only *Apes,* then in all probability our *Ape* may be a *Pygmie ;* a fort of *Animal* fo much refembling *Man ,* that both the Antients and the Moderns have reputed it to be a *Puny Race* of Mankind, call'd to this day, *Homo Sylveftris,* The *Wild Man ; Orang-Outang,* or a *Man* of the *Woods ;* by the *Africans Quoias Morrou ;* by others *Baris,* or *Barris,* and by the *Portugefe,* the *Salvage.* But obferving that under thefe Names, they defcribe different *Animals ;* for Diftinction-fake, and to avoid Equivocation, I fhall call the Subject, of which I am about to give the *Anatomy,* a *Pygmie,* from its Stature ; which I find to be juft the fame with the Stature of the *Pygmies* of the Antients. *Tulpius* 'tis true, and *Bontius,* and *Dapper* do call it, *Satyrus.* And tho' I am of Opinion, that the *Satyrs* of the Antients were of the *Ape,* or rather *Monkey*-kind ; yet for the Reafons alledged in the following *Effay,* I cannot think our *Animal* a *Satyr.* The *Baris* or *Barris,* which they defcribe to be much taller than our *Animal,* probably may be what we call a *Drill.* But I muft confefs, there is fo great Confufion in the Defcription of this fort of Creature, which I find is a very large Family (there being numerous *Species* of them) that in Tranfcribing the Authors that have wrote about them, 'tis almoft impoffible but to make miftakes ; from the want of their well diftinguifhing them. I fhall endeavour therefore in my Account of this, fo to

B   difcridiſcriminate it, that it may be eaſily known again, where-ever 'tis met with. Not that I think in a ſingle Obſervation I can be ſo exact, but that I may be liable to make Errors my ſelf, how careful ſoever I have been.

I will not urge any thing more here, why I call it a *Pygmie* : 'Tis neceſſary to give it a Name ; and if what I offer in the enſuing *Eſſay*, does not ſufficiently Account for the *Denomination*, I leave it to others to give it one more proper. What I ſhall moſt of all aim at in the following Diſcourſe, will be to give as particular an Account as I can, of the formation and ſtructure of all the Parts of this wonderful *Animal* ; and to make a *Comparative* Survey of them, with the ſame Parts in a *Humane Body*, as likewiſe in the *Ape* and *Monkey*-kind. For tho' I own it to be of the *Ape* kind, yet, as we ſhall obſerve, in the *Organization* of abundance of its Parts, it more approaches to the Structure of the ſame in *Men :* But where it differs from a *Man*, there it reſembles plainly the Common *Ape*, more than any other *Animal*.

And tho' I may ſeem too tedious in diſcourſing ſo long upon a ſingle ſubject, yet I have this to offer, that if we had an accurate and particular *Hiſtory* of any *one Species* of *Animal*, it might in a great meaſure ſerve for the *whole kind*. Wherein they differ, might eaſily be taken notice of, and there would be no need of repeating any thing, wherein they all agreed. So formerly diſſecting a Young *Lion* and a *Cat* at the ſame time, I wondred to find ſo very great Reſemblance of all the Parts, both in the one and the other ; that the *Anatomy* of the one might ſerve for the other, allowing for the Magnitude of the Parts, with very little other alteration : And not only for this, but for ſeveral other *Animals*, that belong to the ſame Family. I could have wiſhed I had had the like Opportunity, when I was diſſecting our *Pygmie*, of comparing the ſame Parts with thoſe of an *Ape* and a *Monkey :* For want of it, I have referred all along to the Accounts given us of the *Anatomy* of theſe Creatures by other Authors ; which, tho' it renders my Diſcourſe more prolix, yet I thought it would not be unacceptable to the Curious. But I ſhall take care to draw up in a ſhorter view, wherein our *Pygmie* more reſembled a *Man*, than an *Ape* and *Monkey*,and wherein it differ'd.

Now notwithſtanding our*Pygmie* does ſo much reſemble a *Man* in many of its Parts, more than any of the *Ape-kind*, or any other *Animal* in the World that I know of : Yet by no means do I look upon it as the Product of a *mixt* Generation ; 'tis a *Brute-Animal ſui generis*, and a particular *Species* of *Ape*. For when I was diſſecting it, ſome Sea-Captains and Merchants who came to my Houſe to ſee it, aſſured me, that they had ſeen a great many of them in *Borneo*, *Sumatra*, and other Parts, tho' this was brought from *Angola* in *Africa* ; but was firſt taken a great deal higher up in the Country, and in Company with it there was a *Female* of the ſame kind.

I ſhall have hereafter occaſion to make my Remarks on ſeveral Particulars, relating to it's way of Living , it's Sagacity, Actions, and the like.

like. I fhall now therefore firft of all defcribe its *outward* fhape and figure; then look *within*, and obferve the *Mechanifm* there. But meeting with a *Text* in *Ariftotle*, wherein he gives a general Defcription of the *Ape-kind*, I think it not amifs to Tranfcribe it; and by Commenting upon it, to fhew wherein our prefent Subject agrees with or differs from it; and what I have befides to Remark, I fhall afterwards take notice of, and then proceed to the *Anatomy* of the *Inward* Parts.

Ariftotle's (1) Text is this, which I fhall give with *Jul. Cæf. Scaliger's Latin* Tranflation: And as you may obferve by the Letters of Reference, I have rendred each *Paragraph* into *Englifb*, adding my Obfervations thereon.

(*a*) ΕΝια δ τῶ ζώων ἐπαμφοτερίζει τὴν φύσιν, τῷ τε ἀνθρώπῳ κὴ τοῖς τετράποσιν, οἷον πίθηκοι κὴ κῆβοι, κὴ κυνοκέφαλοι. (*b*) Ἔςι δ᾽ ὁ μὲν κῆβος, πίθηκος ἔχων οὐράν. (*c*) Καὶ οἱ κυνοκέφαλοι δ τὴν αὐτὴν ἔχεσι μορφὴν τοῖς πιθήκοις, πλὴν μεῖζονές τ᾽ εἰσι, κὴ ἰσχυρότεροι, κὴ τὰ πρόσωπα ἔχοντες κυνοειδέςερα. Ἐπὶ δ ἀγριώτερά τε τὰ ἤθη, κὴ τὰς ὀδόντας ἔχεσι κυνοειδεςέρους κὴ ἰσχυρότέρους. (*d*) Οἱ δὲ πίθηκοι, δασεῖς μὲν εἰσι τὰ πρανῆ, ὡς ὄντες τετράποδες· κὴ τὰ ὕπτια δ ὡσαύτως, ὡς ὄντες ἀνθρωποειδεῖς. Τοῦτο γὰρ ἐπὶ τῶ ἀνθρώπων ἐναντίως ἔχει κὴ ἐπὶ τῶ τετραπόδων, καθάπερ ἐλέχθη πρότερον. Πλὴν ἥ τε θρὶξ παχεῖα, κὴ δασεῖς ἐπ᾽ ἀμφότερα σφόδρα εἰσιν οἱ πίθηκοι. (*e*) Τὸ δ πρόσωπον, ἔχει πολλὰς ὁμοιότητας τῷ τοῦ ἀνθρώπου. Καὶ γὰρ μυκτῆρας, κὴ ὦτα παραπλήσια ἔχει· Καὶ ὀδόντας, ὥσπερ ὁ ἄνθρωπος, κὴ τὰς προσθίους κὴ τὰς γομφίους. (*f*) Ἔπι δ βλεφαρίδας, τῶ ἄλλων τετραπόδων ἐκ ἐπ᾽ ἀμφότερα ἐχόντων, οὗτ᾽ ἔχει μὲν, λεπτὰς δ σφόδρα, κὴ μᾶλλον τὰς κάτω, κὴ μικρὰς πάμπαν· τὰ γὰρ ἄλλα τετράποδα ταύτας ἐκ ἔχει. (*g*) Ἔχει δ ἐν τῷ ςήθει δύο θηλὰς μαςῶν μικρῶν. (*h*) Ἔχει δὴ κὴ βραχίονας, ὥσπερ ἄνθρωπος πλὴν δασεῖς κὴ ἰγνύας κὴ τέπυς κὴ τὰ σκέλη ὥσπερ ἄνθρωπος, ταῖς περιφερείας πρὸς

(*a*) INter hominem, quadrupedúmque génus natura quædam media, atque utrique communis eft: Quales, fimia, cebus, caniceps. (*b*) Eft autem cebus fimia caudata. (*c*) Caniceps communem cum fimia formam habet : nifi quod & major & robuftior eft : faciémque habet canine propiorem. Tum moribus exiftunt efferatioribus. Dentes quoque caniniores, atque firmiores. (*d*) Simiæ partes quæ cœlum fpectant, ut pilofæ funt : Proptereà quòd quadrupedum generi afcribuntur : Ita quæ ad terram devergunt quoque : quia hominis fpeciem referunt. Nam in homine, & quadrupedibus hoc contrario fe habere modò fupra dictum eft. Cæterùm fimiis craffus pilus, ac prædenfus utraque in parte eft. (*e*) Ejus verò facies multis modis humanæ fimilis: Quippe tum nares, tum auriculæ : Item dentes tam primores, quàm maxillares funt propemodum tales, quales & homini. (*f*) Quinetiam quadrupedes cæteræ cùm in utraque gena nentiquam palpebras habeant : ipfa habet, fed tennes admodum : tenuiores verò inferiores, atque perpufillos : quibus carent quadrupedes aliæ. (*g*) Et funt in pectore papillæ duæ parvarum mammarum. (*h*) Ad hæc,

(1) *Ariftot. Hift. de Animal. lib.* 2. *cap.* 13. *Ex Edit. Scaliger. cum fuo Com.* p. 197, &c.

B 2   ἀλλήλας.

ἀλλήλας ἀμφοτέρων τῆ κώλων. (i) Πρὸς ἢ τέτοις, χεῖρας ἢ δακτύλες ἢ ὄνυχας ὁμοίως τῷ ἀνθρώπῳ· πλὴν πάντα ταῦτα ἐπὶ τὸ θηριωδέςερον. (k) Ἰδίους δὲ τὰς πόδας εἰσὶ γὰρ οἷον χεῖρες μεγάλαι. Καὶ οἱ δάκτυλοι, ὥσπερ οἱ τῶν χειρῶν· ὁ μέσος, μακρότατος· ἢ τὸ κάτω τῦ ποδὸς, χειρὶ ὅμοιον πλὴν ἐπὶ τὸ μῆκος τῆς χειρὸς ἐπὶ τὰ ἔσχατα τεῖνον, καθάπερ θέναρ. (l) Τῦτο δὲ ἐπ᾽ ἄκρον σκληρότερον, κακῶς, ἢ ἀμυδρῶς μιμούμενον πτέρναν. Κέχρηται δὲ τοῖς ποσὶν ἐπ᾽ ἀμφω, ἢ ὡς χερσὶ, ἢ ὡς ποσὶ, ἢ συγκάμπτει ὥσπερ χεῖρας. (m) Ἔχει δὲ τὸν ἀγκῶνα καὶ τὸν μηρὸν βραχεῖς, ὥσπερ πρὸς τὸν βραχίονα καὶ τὴν κνήμην. (n) Ὀμφαλὸν δ᾽ ἐξέχοντα μὲν οὐκ ἔχει, σκληρὸν δὲ καὶ κατὰ τὸν τόπον τῦτον τῦ ὀμφαλῦ. (o) Τὰ δ᾽ ἄνω τῆ κάτω πολὺ μείζονα ἔχει, ὥσπερ τὰ τετράποδα. Σχεδὸν γὰρ, ὥσπερ πέντε πρὸς τρία θῇ, ἢ διά τε ταῦτα, καὶ διὰ τὸ τὰς πόδας ἔχειν ὁμοίας χερσὶ, καὶ ὡσπερανεὶ συγκειμένους ἐκ χειρὸς καὶ ποδός· οὐκ μὲν ποδός, κατὰ τὸ τῆς πτέρνης ἔσχατον· ἐκ δὲ χειρὸς, τἆλλα μέρη. Καὶ γὰρ οἱ δάκτυλοι ἔχουσι τὸ καλούμενον θέναρ. (p) Διατελεῖ δὲ τ πλείω χρόνον τετράπεζαι ἢ μᾶλλον ἢ ὀρθή. (q) Καὶ ὅτε ἰσχία ἔχει ὡς τετράπεζαν ὂν, ὅτε κέρκον ὡς δίπους, πλὴν μικρὰν τὸ ὅλον, ὅσον σημεῖα χάριν. (r) Ἔχει δὲ καὶ τὸ αἰδοῖον ἡ θήλεια ὅμοιον γυναικός· ὁ δ᾽ ἄῤῥων, κυνωδέστερον ἢ ἀνθρώπου. (s) Οἱ δὲ κῆβοι, καθάπερ εἴρηται πρότερον, ἔχεσι κέρκον· τὰ δ᾽ εἰς τὰ διηρθρωμένα, ὅμοια ἔχεσιν ἀνθρώπῳ πάντα τὰ τοιαῦτα.

hominis brachia, nisi hirta essent. Quæ etiam sicut & crura hominis modo inflectat. Nam & horum, & illorum curvaturas inter se habet contrarias. (i) Tum manus, digitos, ungues, quasi humanos. Verùm hæc omnia ferinam ad naturam potiùs vergunt. (k) Suus quidam modus pedibus, ac peculiaris. Etenim quasi manus quædam magnæ sunt. Quippe & digiti in iis, velnti manuum, medio longissimo. Et planta manni similis, quanquam porrectior ad extremum usque, sicuti vola. (l) Cujus postremum callosus est: inepta, atque inexplanata calcanei similitudine. Pedum usus, & pro manibus, & pro pedibus: flectit enim eos manuum modo. (m) Superior brachij pars, & coxa, breves: si ad ulnæ, & tibiæ magnitudinem referantur. (n) Umbilicus non prominet: sed durum quiddam ibi invenies. (o) Superæ partes inferis majores: quasi si quinarium cum ternario conferas. Hoc autem tum ex quadrupedum natura: tum propterea quòd pedes & manibus similes habet, & quasi ex pedum, manuúmque constitutione compositos. Nam calcanei postrema pedem, cæteræ partes manum repræsentant. Habent enim digiti id, quod volam appellamus. (p) Quadrupedis habitu frequentiore est. (q) Proque eo nates non habet: neque caudam, quoniam bipes. Sed perpusillam omnino illam, & notæ tantùm gratia. (r) Fœminæ genitale muliebri specie est: naribus canina potius, quàm humana. (s) Cebi, sicuti diximus, caudati sunt. Universo generi viscera similia humanis.

(a) Arist.

(*a*) Ariſt. *Some Animals are of an intermediate Nature, between a Man and Quadrupeds, as* Apes, *the* Cebi, *and* Cynocephali.

᾽Επαμφοτερίζει τῃ φύσιν. *Theodorus Gaza* thus renders this Paſſage : *Sunt quæ natura ancipite, partim hominem, partim Quadrupedem imitentur, ſicut ſimiæ,* &c. Not that an Ape is part a Man, and part a Quadruped ; *inter Hominem & non Hominem non datur medium* ; The Terms being contradictory, one muſt be falſe. The Philoſopher's meaning muſt therefore be, that in the formation of the Parts of the Body, the *Ape,* the *Cebus,* and *Cynocephalus,* are intermediate Species between a *Man* and other *Quadrupeds,* having ſeveral Parts of the Body formed like *Brutes* ; others more reſembling thoſe of *Men.* (2) *Scaliger,* a little after, hath this Remark ; " Ad eum namque modum ſummus Opifex Rerum ſeriem " concatenavit a Planta ad Hominem ; ut quaſi ſine ullo cohæreant in- " tervallo, ſic ζωόφυτα cum Plantis Bruta conjungunt ; ſic cum homine " ſimia Quadrupedes. Itaque in hominis quoque ſpecie inveniamus " Divinos, Humanos, feros. This *Climax* or *Gradation* can't but be taken notice of, by any that are curious in obſerving the Wonders of the *Creation* ; and the more he obſerves it, the more venerable *Idea's* 'twill give him of the great *Creator* ; and it would be the Perfection of *Natural Hiſtory,* could it be attained, to enumerate and remark all the different *Species,* and their *Gradual Perfections* from one to another. Thus in the *Ape* and *Monkey*-kind, *Ariſtotle's Cebus* I look upon to be a degree above his *Cynocephalus* ; and his *Pithecus* or *Ape* above his *Cebus,* and our *Pygmie* a higher degree above any of them, we yet know, and more reſembling a *Man :* But at the ſame-time I take him to be wholly a *Brute,* tho' in the formation of the Body, and in the *Senſitive* or *Brutal* Soul, it may be, more reſembling a Man, than any other *Animal* ; ſo that in this *Chain* of the *Creation,* as an intermediate Link between an *Ape* and a *Man,* I would place our *Pygmie.*[*]

Πίθηκος, &c. The Philoſopher here does not enumerate all the ſeveral *Species* that are contained under the *Ape* and *Monkey*-kind ; they are a very numerous and a large *Claſſis* of Animals. *Scaliger* upon the Place mentions ſeveral he had obſerved of both kinds ; and all our *Zoographers,* and moſt Journals of Travels give a Deſcription of a great many ſorts of them. But for want of well diſtinguiſhing them, and ranging them into a Methodical *Series,* their *Hiſtory* as yet is very confuſed and perplext. Mr. *Ray* (3) places theſe Animals under this general *Title, Animalia Pede unguiculato multifido,* πλατυώνυχα & ανθρωπόμορφα. 'Tis call'd *Pithecus,* παρὰ τὸ πείθεθαι ὑφ᾽ ἡμῶν, *quia facile ab homine perſuadeatur* ; and oftentimes this word is taken as a *Genus* which includes the whole ; when ſtrictly taken, it ſignifies an *Ape* without a Tail, and in *Latin* is call'd *Simia* ; that which hath a Tail is call'd *Cercopithecus,* in *Engliſh* a *Monkey.* Thus (4) *Martial.*

(2) Scaliger *ibid. in Com,* pag. 201. (3) Raij *Synopſis Animal.* pag. 148. (4) Martial. *Epigram. lib.* 14. *Epigr.* 202.

Callidus

Callidus emiſſas eludere Simius Haſtas,
Si mihi Cauda foret, Cercopithecus eram.

(*b*) Ariſt. *The Cebus is an Ape having a Tail.*

(5) *Conradus Geſner* thinks, that this *Cebus* of *Ariſtotle*, which he deſcribes only as having a Tail, muſt be the *Cercopithecus* or Common
Monkey, ſince he mentions not the *Cebus* any where elſe, and the *Cercopi-*
thecus no where. (6) *Harduinus*, in his Notes on *Pliny*, adviſes not to
miſtake the *Cepus* in *Pliny*, for the *Cebus* in *Ariſtotle*. (6) *Pliny's* words are
theſe ; *Pompeii Magni primum Ludi oſtenderunt Chama, quem Galli Ru-*
fium vocabant, Effigie Lupi, Pardorum maculis. Iidem ex Æthiopia quas
vocant κῆπυς, quarum Pedes poſteriores, Pedibus humanis & cruribus, pri-
ores manibus fuere ſimiles, hoc Animal poſtea Roma non vidit. And therefore becauſe it was ſo uncommon as to be ſeen at *Rome* but once, it
could not be the common *Monkey*. (7) *Strabo*, out of *Artemidorus*,
deſcribes the *Cepus* thus : γίγνονται δὲ φησι ᵏὖ σφίγγες, ᵏὖ κυνοκέφαλοι, ᵏὖ κῆ-
φοι, λέοντες μὲν πρόσωπον ἔχοντες, τὸ δὲ λοιπὸν σῶμα πάνθηρος, μέγεθος δὲ
δορκάδος. That the *Cepus* hath the Face of a *Lion*, the reſt of the
Body like a *Panther*, and is of the bigneſs of a *Dorcas* or *Roe-Buck*.
(8) *Diodorus Siculus* hath much the ſame Deſcription, ὁ δὲ λεγόμενος
κῆπος, ἀνόμαςαι μὲν ᵃπὸ τῆς περὶ ὅλον τὸν ὄγκον ὡραίας, ᵏὖ περφανῶς ἡλικίας.
Τὸ δὲ πρόσωπον ἔχων ὅμοιον λέοντι, τὸ λοιπὸν σῶμα φέρει πάνθηρι παρα-
πλήσιον, πλὴν τῷ μεγέθει, ὁ παρισοῦται δορκάδι. Which *Laurentius Ro-*
domannus thus renders. " Cepus, *i. e.* Hortus (quem vocant) à totius
" Corporis decore & ſtaturæ venuſtate nomen accepit, facie Leonem imi-
" tatur, & reliquo Pantheram, præter magnitudinem, qua Dorcadi par
" eſt. (9) *Ælian* hath given a Deſcription of the ſame Animal from
Pythagoras, from whom, 'tis thought, it firſt received this Name ; and
he is more particular. His Account, tho' ſomewhat long, I will give
in *P. Gillius's* Tranſlation, becauſe I am apt to think this *Animal* is ſtill
in being. " Terrenum quoddam Animal Pythagoras ſcribit ſecundùm
" Mare Rubrum procreari & Cepum, hoc eſt Hortum appoſitè idcircò
" nominari, quòd tanquam Hortus variis coloribus diſtinguatur. Cùm
" exiſtit confirmata ætate, pari magnitudine eſt cum Herythrienſibus
" Canibus. Jam porro ejus Colorum varietatem, ſicut ille ſcribit, ani-
" mus nobis eſt explicare. Ejus caput & poſticas partes ad caudam uſque
" prorſus valde igneo colore ſunt, tum aurei quidam Pili diſſeminati
" ſpectantur, tum album roſtrum, inde ad Collum aureæ vittæ pertinent,
" Colli inferiores partes ad Pectus, & anteriores Pedes omnino albi,
" Mammæ duæ manum implentes cæruleo colore viſuntur, venter candi-
" dus, Pedes poſteriores nigri ſunt, Roſtri formæ Cynocephalo recte

---

(5) *Hiſt. de Quadruped.* l. 1. p. 857. (6) Plinij *Hiſt. Nat.* lib. 8. cap. 19. *cum Interpret. & Notis*
Jo. Harduini, p. 167. (7) *Geograph.* lib. 16. p. 533. (8) Diodor. Sicul. *Biblioth. Hiſt.* l. 3. p.m. 168.
(9) Ælian. *de Animal.* lib. 17. cap. 8. p. 474.

" com-

" comparari poteſt. The *Cepus* therefore of *Pliny, Strabo, Diodorus Siculus,* and *Ælian,* in all probability muſt be different from the *Cebus* of *Ariſtotle.* *Joh. Caius* our Country-man ſent *Geſner* a Deſcription of a *Mamomet* or *Marmoſet* he had obſerved, which *Geſner* thinks might be a ſort of *Cepus* ; but the Colours were different, as likewiſe the Magnitude.

(c) Ariſt. *The Cynocephali have the ſame ſhape with* Monkeys, *but they are bigger, and ſtronger, and they have a Face liker a Dog's, and are of a fiercer Nature, and they have Teeth liker a Dog's, and ſtronger.*

I ſhall have occaſion to Diſcourſe of theſe *Cynocephali* in the enſuing *Eſſay.* For tho' the Philoſopher makes them only a ſort of *Ape* or *Monkey,* yet there have been thoſe, that would impoſe them on the World for a Race of *Men* ; and by (10) *Ælian* they are call'd ἄνθρωποι κυνοπρόσωποι ; tho' (11) *Galen* tells us, they are much leſs like a *Man,* than an *Ape* is: For they can ſcarce ſtand upright, much leſs walk or run ſo. (12) *Philoſtorgius* mentions the *Aegopithecus,* the *Arctopithecus,* the *Leontopithecus,* as well as the *Cynocephalus,* and then adds, καὶ ἄλλαις πολλῶν ζωῶν εἰδέαις τῆς πιθηκείας μορφῆς ἐπιμιγνυμένης. That there is the *Goat-Ape,* the *Bear-Ape,* the *Lion-Ape,* the *Dog-Ape* ; and that the *Ape-kind* have a reſemblance to a great many other *Animals* ; ſo large and numerous is this *Claſſis* of Animals, that perhaps there is none that is more ; and that are ſo different from one another. The *fierceneſs* of the *Cynocephali* is taken notice of by all ; our *Pygmie* was quite of another temper, the moſt gentle and loving Creature that could be. Thoſe that he knew a Ship-board he would come and embrace with the greateſt tenderneſs, opening their Boſoms, and claſping his Hands about them ; and as I was informed, tho' there were *Monkeys* aboard, yet 'twas obſerved he would never aſſociate with them, and as if nothing a-kin to them, would always avoid their Company. The *Teeth* of the *Cynocephali* are like a *Dog's* ; thoſe of our *Pygmie* exactly reſembled a *Man's,* as I ſhall ſhew in the *Oſteology.*

(d) Ariſt. *Apes are hairy on their Backs, as they are Quadrupeds, and on their Bellies, as they are like Men : For in a Man and a Beaſt this hairineſs is quite contrary, as was ſaid before. So that Apes are very hairy in both Places, their Hair being ſtrong or courſe, and thick ſet.*

The Place that *Ariſtotle* refers to, is this. (13) Ἔστι δὲ τῶν μὲν ἄλλων ζώων τῷ ἔχοντων τρίχας, τὰ πρανῆ δασύτερα, τὰ δ᾽ ὕπτια, ἢ λεῖα πάμπαν, ἢ δασέα ἧττον. ὁ δ᾽ ἄνθρωπος τουναντίον. i. e. *That in Brutes the Back or upper Parts are more hairy, the Belly or under Parts either ſmooth or leſs hairy : In a Man is obſerved the contrary.* But in our *Pygmie* we obſerved it different ; for here all behind from the Head downwards, 'twas very hairy, and the Hair ſo thick, that it covered the Skin almoſt from being ſeen.

(10) Ælian. *Hiſt. de Anim.* lib. 10. cap. 26. *in Edit.* P. Gillij. *in aliis cap.* 25. (11) Galen. *de Adminiſtr. Anat.* l. 1. cap. 2. (12) Philoſtorgij *Hiſt. Eccleſiaſt.* lib. 3. cap. 11. p. 41. (13) Ariſt. *Hiſt. de Animal.* lib. 2. c. 5. p. 160. *Edit.* Scalig.

But

But in all the Parts before, the Hair was much thinner, and the Skin every where appeared, and in ſome places 'twas almoſt bare. Nature therefore has cloathed it with Hair, as a Brute, to defend it from the Injuries of the Weather; and when it goes on all four, as a *Quadruped,* it ſeems all hairy : When it goes erect, as a *Biped,* it appears before leſs hairy, and more like a *Man.* After our *Pygmie* was taken, and a little uſed to wear Cloaths, it was fond enough of them; and what it could not put on himſelf, it would bring in his Hands to ſome of the Company to help him to put on. It would lie in a Bed, place his Head on the Pillow, and pull the Cloaths over him, as a Man would do; but was ſo careleſs, and ſo very a Brute, as to do all Nature's Occaſions there. It was very full of Lice when it came under my Hands, which it may be it got on Ship-board, for they were exactly like thoſe on Humane Bodies. (14) *Seignior Redi* obſerves in moſt Animals a particular ſort of Louſe, and gives the Figures of a great many.

The Hair of our *Pygmie* or *Wild Man* was of a Coal-black colour, and ſtrait; and much more reſembling the Hair of *Men* than the Furr of Brutes : For in the Furr of Brutes, beſides the longer Hair, there is uſually a finer and ſhorter *Pile* intermixt : Here 'twas all of a kind; only about the *Pubis* the hair was greyiſh, ſeemed longer, and ſomewhat different; ſo on the upper Lip and Chin, there were greyiſh hairs like a *Beard :* And I was told by the Owners, that once it held the Baſon it's ſelf, to be trimmed. The Face, Hands, and Soles of the Feet were bare and without Hair, and ſo was moſt part of the Forehead : But down the ſides of the Face 'twas very hairy; the hairs there being about an Inch and half long, and longer than in moſt Parts of the Body beſides. The tendency of the Hair of all the Body was downwards; but only from the Wriſts to the Elbow 'twas upwards; ſo that at the Elbow the Hair of the Shoulder and the Arm ran contrary to one another. Now in *Quadrupeds* the Hair in the fore-limbs have uſually the ſame Inclination downwards, and it being here different, it ſuggeſted an Argument to me, as if Nature did deſign it as a *Biped.* But we will lay no more ſtreſs upon it than it will bear : The Hair on the back-ſide of the Hands did run tranſverſe, inclining to the outſide of the Hands; and thoſe of the hinder ſides of the Thighs were tranſverſe likewiſe.

Man, tho' not ſo hairy as *Brutes,* and (as *Ariſtotle* obſerves) more hairy before, than behind; yet if expoſed to the hardſhips of the Weather, like them; no doubt, but he would become hairy on the Body likewiſe; which might poſſibly be the Caſe of *Nebuchadnezzar.* (15) And very Remarkable is that Story of *Peter Serrano* a *Spaniard,* who was caſtaway, and eſcaped to a Deſart Iſland, which from him afterwards received it's Name, as 'tis related by the *Inca Garcilaſſo de la Vega.* (16) For having with the greateſt difficulty ſuſtained a miſerable Life for three

---

(14) Franc. Redi *Experimenta circa generat. Inſector.* (15) Daniel. *Cap.* 4. 33. (16) Royal Commentaries of *Peru.* lib. 1. cap. 3.

Years,

Years, " The Hairs of his Body grew in that manner, that he was co-
" vered all over with Briſtles ; the hair of his Head and Beard reach-
" ing to his Waſte, that he appeared like ſome Wild or Savage Crea-
" ture.

(e) Ariſt. *Their Face hath many Reſemblances to a Man's, for they have
Noſtrils and Ears alike ; and Teeth like a Man's, both the Fore-teeth 'and
the Grinders.*

Pliny (17) ſeems to have reſpeCt to this *Text* of *Ariſtotle,* and what
follows, where he tells us, " Nam ſimiarum genera perfeCtam Hominis
" imitationem continent, facie, Naribus, Auribus, Palpebris, quas ſolæ
" Quadrupedum in inferiore habent Genâ. Jam Mammas in PeCtore ,
" Brachia & Crura in contrarium ſimilitèr flexa. In manibus, ungues,
" digitos , longioreinque medium. Pedibus paulùm differunt , ſunt
" enim, ut manus, præ̀longi, ſed veſtigium Palmæ ſimile faciunt. Pol-
" lex quoque his & Articuli, ut homini ; ac præter Genitale, & hoc in
" maribus tantùm. Viſcera etiam interiora omnia ad exemplar. We
will compare both their Accounts, with our *Pygmie* ; and obſerve wherein they agree or differ from us.

As for the *Face* of our *Pygmie,* it was liker a *Man's,* than *Ape's* and
Monkeys Faces are : For it's *Forehead* was larger, and more globous, and
the upper and lower *Jaw* not ſo long or prominent, and more ſpread ;
and it's *Head* more than as big again as either of theirs : But why the
Philoſopher, after his general Aſſertion of the likeneſs of the *Face* of an
Ape to that of a *Man's,* ſhould firſt of all inſtance in the *Noſe,* which
is ſo much different, may ſeem ſtrange : Since in a *Man* the *Noſe* is protuberant and riſing, jutting out much beyond the whole ſurface, and
herein 'tis altogether unlike to that of Brutes, and the *Ape*-kind too.
'Tis not therefore on this account that the Compariſon is made. But I
rather think, his meaning muſt be, that an *Ape's Noſe* is like a *Man's,*
in that it is not extended to the length of the *Roſtrum,* or upper *Jaw,*
as in Dogs and other Brutes, but reaches only to the upper *Lip. à ſimis
Naribus,* or this flatneſs of the *Noſe,* moſt do derive the word *Simia ;*
tho' others, as *Voſſius,* would have it, *quaſi mimia* à μιμεἴϑαι, *imitari,*
from mimicking. But *Scaliger* will not allow it. *Dicitur autem Simia*
(ſaith he) *non ab Imitatione, ut Grammatici imperiti, ſed à ſimitate.*

The *Noſe* of our *Pygmie* was flat like an *Ape's,* not protuberant as a
Man's ; and on the outſide of each *Noſtril* there was a little ſlit turning
upwards, as in *Apes.* 'Tis obſerved of the *Indian Blacks,* that their
Noſe is much flatter than the *Europeans* ; which may be thought rather
Natural to that Nation, than occaſioned (as ſome would make us be-
lieve) by the Mother's tying the Infant to her Back, and ſo when at
Work bruiſing and flatting it againſt her Shoulders ; becauſe 'tis ſo uni-
verſal in them all.

(17) *Natur. Hiſt.* lib. 11. cap. 44. p. m. 593.

C

As

As to the *Ears*, none could more refemble thofe of a *Man*, than our *Pygmie's*; both as to the largenefs, colour, fhape, and ftructure: Here I obferved the *Helix, Ant-Helix, Concha, Alvearium, Tragus, Anti-tragus,* and *Lobus*; only the *Cartilage* was very fine and thin, and the *Ears* did not lye fo flat to the *Head*, as they do in a *Man*. But that may be from the Cuftom of binding our Heads, when Infants.

The *Teeth* of our *Pygmie* refembled a *Man's*, more than do thofe of *Apes* and *Monkeys*; as I fhall fhew in the *Ofteology*.

(*f*) Arift. *And whereas other Quadrupeds have not Hair on both Eye-lids, thefe have; But 'tis very fine, efpecially that on the lower Eye-lid, and very fmall. But other Quadrupeds have none there.*

In our *Pygmie* the *Cilia* or Hair of both Eye-lids appeared very fair and plain, but not fo large as in *Men*. The *Supercilia* or Hair of the *Eye-brows*, feem'd to be rubb'd off; which might be occafioned by the jutting out of the *Cranium* in that place, more than in *Men* : Which is a Provident Provifion of *Nature*, for the better fafeguard of the *Eyes*, and their defence from the Injuries they might otherwife receive in the Woods. But the *Philofopher's* Affertion, that no *Quadruped* hath *Hair* on the *under Eye-lid* befides *Man* but the *Ape*-kind, I cannot juftifie; or I do not take his meaning aright: Tho' he has much the fame Opinion a little before. (18) Where he tells us, Καὶ Ϭλεϼϗρϼίδας ὁ μὲν ἄνθρωπος ἐπ' ἀμφω ἔχει, ἡ ἐν μαχάλαις ἔχει τϼίχας, ἡ ὑπὶ τῆς ἥϭης. Τῶν δ᾽ ἄλλων ὄδὲν ὅτε τότων ὀδέτεϼον, ὅτε τὴν κάτωθεν Ϭλεφαϼίδα; ἀλλὰ κάτωθεν τῆ Ϭλεφάϼϼ ἐνίοις μαναὶ τϼίχες πϼύκασιν. Which *Scaliger* thus renders: *Ac Palpebras homo utràque in Genâ habet tum & in Alis, & in Pube Pilos. Cæteræ Animantes neque in his locis, neque in Genâ inferiore : Sed fub Genam & paucos & paucæ.* Our *Pygmie* had Hair in the *Arm-pits*, and that in the *Pubis* feemed fomewhat different from what grew on the reft of the Body; being not fo ftrait, but fomewhat curled; and greyifh, not black. But I muft here Remark, that *Pliny* ufes the words *Palpebræ* and *Gena*, in his Tranflating this Text of *Ariftotle*, different from what commonly they fignifie now. For by *Palpebræ* he means, what *Ariftotle* and *Hippocrates* call Ϭλεϼϗϼίδας, *i. e.* the *Hair* on the Rim of the *Eye-lids, à palpitatione*; and *Feftus* calls *Cilia, quia oculos celent & tueantur*: And by *Gena*, he underftands the *Eye-lid*; as appears from that Paffage of *Pliny* I have juft now quoted, , *Palpebris quas folæ Quadrupedum in inferiore habent Genâ.* And fo *Scaliger* ufes thefe words in this Tranflation of *Ariftotle* : And he makes *Cilium* to fignifie, *Summum Genæ ambitum*, and not the Hair there.

(*g*) Arift. *They have two Teats or Nipples of fmall Breafts on the Ster-num.*

(18) *Hift. Animal.* lib. 2. p. m. 161.

The

The *Philosopher* here obferves, That the *Ape-kind*, common with *Humane*, have the *Mammæ* on the *Sternum* or Breaft, which is different from *Brutes*. And tho' the *Elephant* herein feems fomewhat alike, yet he makes this diftinction, (19) ὁ δὲ ἐλέφας ἔχει μὲν μαςὲς δύο, ἀλλ' ὐκ ἐι τῷ ςήθει, ἀλλὰ πρὸς τῷ ςήθει. *Juxta Pectus potius, quàm in Pectore*, as *Scaliger* renders it ; or as *Theodorus Gaza, non in Pectore, fed paulò citra*. And a little after, (20) he more particularly expreffes himfelf, κ̀, γὸ ὁ ἐλέφας ἔχει τὰς μαςὲς δύο περὶ τὰς μαχάλας. *Sub Armis*, as *Gaza* renders it ; *ad Axillas*, as *Scaliger*, where he further tells us, That the *Male* as well as *Female Elephant* have thefe *Teats* ; but they are very fmall, in refpect of the Bulk of it's Body, and fo placed that fide-ways, you can't fee them. The *Bear* (he adds) hath four *Teats* ; *Sheep* have but two, and thofe between the hinder Legs ; *Cows* have four *Teats* there. Other *Animals* (he faith) have thefe *Teats* in the middle of the *Belly*, and ufually more numerous ; as the *Dog* and *Swine-kind* : But the *Panther* hath but four in the Belly : The *Camel* hath two *Mammæ* there, and four *Teats*, as a *Cow* ; and a *Lionefs* but two there.

But *Apes* and *Monkeys* have their *Teats* upon the *Breaft* ; as *Women* have ; and (21) *Albertus Magnus* gives this Reafon for it, *Mammillas autem habet in Pectore ficut Mulier, eò quòd manus dedit ei Natura, quibus ad Pectus poteft elevare partum, ficut Mulier*. Our *Pygmie* was a *Male*, yet here the two *Papillæ* or *Teats* appeared very plain, and were exactly fituated as they are in Men. The *Mammæ* or *Breafts* were fmall and thin, and not protuberant. The *Female Orang-Outang* of (22) *Bontius* is pictured with pendulous large *Breafts*, and they are fo defcribed by (23) *Tulpius*. And (24) *Gaffendus*, in the Life of *Peiresky*, fpeaking of the *Barris*, faith, *Huic Mammæ ad pedis longitudinem*.

(h) Arift. *They have Arms like a Man, but hairy ; and they bend them and the Legs as a Man does ; the flection of the one being contrary to the other.*

The *Shoulder* and *Arm* of our *Pygmie* were very hairy outwards, not fo hairy inwards. The Contratendency of the Hair here, as that of the *Shoulder* pointing downwards, and that of the *Arm* pointing upwards, like *Lucan's Pila minantia Pilis*, I have already noted. This difference I fhall here remark of this fore-limb in our *Pygmie*, as well as in *Apes* and *Monkeys* ; that 'tis longer in them proportionably, than in *Man*. I fhall examine this Part more particularly in the *Myology* and *Ofteology*.

But the Curvature or Flection of the *Arms* and *Legs* in our *Pygmie*, as alfo in *Apes* and *Monkeys*, is juft the fame as in *Man* ; the *Arms* bending forwards, and the *Legs* backwards ; whereas in other *Brutes*, the flection

(19) Arift. *ibid.* p. 151.   (20) Arift. *ibid.* p. 176.   (21) Albert. *de Animal.* lib. 22. p. 224.
(22) Jac. Bontij *Hift. Nat. & Med.* lib. 5. cap. 32. p. 84.   (23) Nic. Tulpij *Obferv. Med.* l. 3. cap. 56.
(24) Gaffend. *de vita Peireskij.* lib. 5. p. m. 170.

C 2   of

of the fore and hinder *Legs* is both the ſame way. *Homini Genua &*
Cubita contraria (ſaith (25) *Pliny*) *item Urſis & ſiniarum generi, ob id*
minime pernicibus. I ſhall examine this Place of *Pliny* in the *Oſteo-*
logy.

(*i*) Ariſt. *Beſides they have Hands, Fingers, and Nails like a Man's,*
but all theſe ſomewhat ruder.

The *Hand* of our *Pygmie* was different from a *Man's*, in that the
Palm was much longer ; ſo the *Thumb* too, was leſs than the other *Fin-*
gers ; whereas in a *Man*, the *Thumb* is uſually thicker than the reſt of the
Fingers : In both theſe reſpects, it more reſembled the *Ape*-kind. But
the *Fingers* of our *Pygmie* being ſo much bigger than thoſe of *Apes* and
Monkeys ; and its *Nails* being broader, and flatter, on both theſe Accounts it was liker a *Man.* *Ungues Clauſulæ Nervorum ſummæ exiſtiman-*
tur (ſaith (26) *Pliny*) *omnibus hi, quibus & digiti : ſed Simiæ imbricati,*
Hominibus lati.

In the *Palms* of the *Hands* of our *Pygmie* were remarkable thoſe *Lines*
which are uſually taken notice of in *Palmeſtry* ; and at the ends of the
Fingers were thoſe *Spiral Lines*, which are uſually in a *Man's.*

(*k*) Ariſt. *The Feet are particular ; for they are like great Hands, and*
the Toes like Fingers ; the middlemoſt being the longeſt : And the Sole of
the Foot like the Palm of the Hand, but more extended, or longer.

Pliny (as I have remark'd) renders this Paſſage thus: *Pedibus paulum*
differunt, ſunt enim, ut manus, prælongi ; ſed veſtigium Palmæ ſimile fa-
ciunt. Now the *Palms* of the *Hands*, and the *Soles* of the *Feet* of our
Pygmie, were equally long, and longer, proportionably, than in *Man* ;
and herein it reſembled more the *Ape*-kind : As it did likewiſe in the
length of the *Toes*, which were as long as the *Fingers*, as alſo in having
the middlemoſt *Toe* longer than the reſt. For in the *Hand* of a *Man*,
the middle Finger is the longeſt, but in the *Foot*, the middle Toe is not.
The *Philoſopher* does very well liken it to a Hand, ſince beſides the length
of the *Toes*, like *Fingers*, it had the *great Toe*, like the *Thumb* ſet off at
a diſtance from the range of the other Toes, as we ſhall ſhew hereafter.

(*l*) Ariſt. *The ſole of the Foot in the hinder part was more callous, ill,*
and odly imitating a Heel: For they uſe their Feet in both Capacities, both
as a Hand and Foot, and bend them like Hands.

In the *Ape*-kind there is a true *Os Calcis*, beſides this *Calloſity.* And in
our *Pygmie* this *Heel-bone* was liker that in a Man, than theirs is. The
Philoſopher in the former *Paragraph* ſhewed what reſemblance this Part
had to a *Humane Hand*, in this, by reaſon of the *Os Calcis*, how 'tis like

(25) Plinij *Nat. Hiſt.* l. 11. cap. 45. p. m. 594.   (26) Plinij *Nat. Hiſt.* lib. 11. cap. 45. p. 594.

a *Foot* ; and then makes an Inference from the different ſtructure of this *Organ*, that it performs the Uſes and Offices of both.

All which is very agreeable to our *Pygmie*. But this *Part*, in the Formation and it's Function too, being liker a *Hand*, than a *Foot* ; for the diſtinguiſhing this ſort of *Animals* from others, I have thought, whether it might not be reckoned and call'd rather *Quadru-manus* than *Quadrupes*, i. e. a *four-handed*, than a *four-footed Animal*.

And as it uſes it's hinder *Feet* upon any occaſion, as *Hands* ; ſo likewiſe I obſerved in our *Pygmie*, that it would make uſe of it's *Hands*, to ſupply the place of *Feet*. But when it went as a *Quadruped* on all four, 'twas awkwardly ; not placing the *Palm* of the *Hand* flat to the Ground, but it walk'd upon it's Knuckles, as I obſerved it to do, when weak, and had not ſtrength enough to ſupport it's Body. So that this *Species* of *Animals* hath the Advantage of making uſe of their *Feet* as *Hands*, and their *Hands* as *Feet*, as there is occaſion.

(*m*) Ariſt. *The* Os Humeri, *and the* Os Femoris *are ſhort, in reſpect of the* Ulna *and* Tibia.

In a *Humane Skeleton*, the *Os Humeri*, and the *Os Femoris* are much longer than the *Ulna* and *Tibia*. For in a *Skeleton* of a Woman I have by me, the *Os Humeri* was Twelve Inches and a half, and the *Os Femoris* Seventeen Inches long ; whereas the *Ulna* was but Nine Inches and three quarters, and the *Tibia* Fourteen Inches long. In our *Pygmie*, the *Os Humeri* was Five Inches and a half, and the *Os Femoris* Five Inches long. The *Ulna* was Five Inches and a half, and the *Tibia* was Four Inches long. Theſe *Bones* in the *Skeleton* of a *Monkey*, were much of the ſame length with our *Pygmie's*, ſo that herein both differ from a *Man*, and our *Pygmie* more reſembles the *Ape-kind*.

(*n*) Ariſt. *They have no prominent Navel, but ſomething hard in this place of the Navel.*

In our *Pygmie* the *Umbilicus* or Navel appeared very fair, and in the exact Place, as 'tis in a Man ; not prominent nor harder, but in all reſpect Natural and alike.

(*o*) Ariſt. *They have the upper Parts much larger than the lower, as being Quadrupeds, almoſt as five to three ; and as upon this account, ſo becauſe they have Feet like Hands, as if they were compounded of a Hand and Foot : Of a Foot, in reſpect of the Heel behind ; and of a Hand, as to the other Parts ; for they have Fingers, and what we call the Palm.*

In Quadrupeds uſually the *Upper* or fore-parts are much larger than the *Lower* or hinder ; and 'tis ſo in the *Ape* and *Monkey*-kind, as the *Philoſopher* Remarks. But in our *Pygmie* I think this Obſervation will not hold. For tho' it was much emaciated, by reaſon of it's long illneſs, ſo that it ſeemed very thin and lank in the *Belly* ; yet behind it look'd ſquare enough, and proportionable as a Man. But the *Orang-Outang* of *Tulpius*

had

had a large ſquob *Belly.* We ſhall preſently give the Dimenſions of all the Parts, as ſoon as we have done with this *Text* of *Ariſtotle.*

We ſhall hereafter farther conſider the ſtructure of the *Foot* in the *Oſteology,* where we ſhall deſcribe the *Os Calcis,* and ſhew how well it performs its Office, when this *Animal* ſtands erect. But ſince *Nature* deſign'd it not always to live on the Ground, but to get it's Prey in the Trees likewiſe, it hath very wiſely formed this Part like a Hand, by which means it can more eaſily climb them ; and when there, ſhift much better by this Contrivance ; as I have ſhewn in my Diſcourſe (27) upon the *Carigueya, ſeu Marſupiale Americanum,* or the *Anatomy* of an *Opoſſum* ; which *Animal* had its hinder *Feet* formed like *Hands.*

(p) Ariſt. *They live moſt of their time as Quadrupeds more than as Bipeds, or erect.*

Our Merchants tell me, when firſt they take *Apes* or *Monkeys,* to learn them to go erect, they uſually tye their Hands behind them. And I am of the *Philoſopher's* Mind, that Naturally they go more on all Four, than erect. But whether 'tis ſo in our *Pygmie,* I do ſuſpect ; ſince walking on it's Knuckles, as our *Pygmie* did, ſeems no Natural Poſture ; and 'tis ſufficiently provided in all reſpects to walk erect.

(q) Ariſt. *As Quadrupeds they have no Buttocks; as Bipeds, no Tails; or but very little, like a ſhew of one.*

Our *Pygmie* had *Buttocks* or *Nates,* as we ſhall ſee in the *Myology,* but not ſo much as in *Man.* The *Os Iſchij* or *Coxendix* was very different, as appears in the *Skeleton,* and as I ſhall deſcribe in the *Oſteology.* Our *Pygmie* had no *Tail,* but an *Os Coxygis,* as is in Man, which outwardly made a little appearance, as in my *Second Figure,* and may be what *Ariſtotle* Remarks. *Scaliger* has this Note upon it : *Caudæ notam ſive veſtiginm animadvertit, quam vix Oculis deprehendas. Tactu tamen ſubeſſe intelligas, quam ſi attractare tentes, promptâ miráque celeritate ſeſe ſubtrahit, ridiculâ indignatione læſum præ ſe fert.*

(r) Ariſt. *The Female hath the Privy-parts, like a Woman; but the Male, more like a Dog's, than a Man's.*

Our Subject was a *Male,* and this Part here was nothing like a *Dog's.* For in the *Penis* of a *Dog* there is a large *Bone,* which is not in the *Ape* and *Monkey-*kind. *Scaliger's* Note here does not make out the Aſſertion : *Caninum Genitale dixit Simij, non temeré ; nodos enim quoſdam deprehendimus : differt autem figurâ Glandis.* I did not obſerve theſe Nodes here ; but of this, more in the *Anatomy* of this Part.

(s) Ariſt. *The* Cebi *(as was ſaid before)* have Tails *: As to the Viſcera they have them all like a Man's.*

(27) *Philoſoph. Tranſact.* Numb. 239.

So *Pliny, Viscera etiam interiora omnia ad Exemplar.* But I find this a great Miftake. For, as we fhall fhew, our *Pygmie,* who comes much nearer to a *Man* in the Structure of the inward Parts, than either *Apes* or *Monkeys,* yet in a great many things is very different ; but where it is fo, there it refembles an *Ape.* But on the other hand, *Albertus Magnus* is much more miftaken, who will not allow any likenefs at all. For fpeaking of an *Ape,* he tells us, (28) *Et ficut in ante habitis diximus, homini in exterioribus fimile exiftens , in nullo fimilitudinem habet cum interioribus hominis, & minùs fere omnibus aliis Beftiis.* Galen (29) is much more in the right, who acknowledges a very great fimilitude between an *Ape* and a *Man,* both in the outward and inward Parts, where he tells us, Καὶ ὁ πίθηκος ἁπάντων τῶν ζώων ὁμοιότατος ἀνθρώπῳ, ἐ σπλάγχνοις, ἐ μυσὶ, ἐ ἀρτηρίαις, ἐ φλεψὶ, ἐ νεύροις, ὅτι ἐ τῇ τῶν ὀςῶν ἰδέᾳ. Διὰ γὸ τὴν τέτων φύσιν ἐπὶ τε δυοῖν βαδίζει σκελοῖν, ἐ τοῖς προϑίοις κώλοις ὥσπερ χερσὶ χρῆται, ἐ ςέρνον πλατύτατον ἁπάντων τῶν τετραπόδων ἔχει, ἐ κλεῖς ὡσαύτας ἀνθρώπῳ, ἐ πρόσωπον ςρογγύλον, ἐ τράχηλον μικρόν. i. e. *An Ape is the moft like a Man of any Quadruped : In the Viscera and the Mufcles, and in the Arteries, and Veins and Nerves; becaufe 'tis fo in the ftructure of the Bones. For 'tis from their make, that it walks on two Legs, and ufes its fore-limbs as Hands. It hath the largeft Breaft of any Quadruped, and Clavicles or Collar-bones like a Man, and a round Face, and a fmall or fhort Neck.*

All which is very agreeable to our *Pygmie,* whom we fhall find more exactly to anfwer this Character, than an *Ape.* And now having compared our *Pygmie* with this general Defcription that *Ariftotle* gives of the *Ape*-kind ; we fhall compare him with himfelf, by taking the different *Dimenfions* of the feveral *Parts,* as well as of the whole *Body* ; and fhall obferve what Proportions they had to one another.

As from the top of the Head, to the heel of the Foot in a ftrait Line, it meafured Twenty fix Inches. The Girth of the Body in the biggeft part about the *Cartilago Enfiformis,* was Sixteen Inches ; over the Loins 'twas Ten Inches about. The Compafs of the Head over the Eyes and Ears, Thirteen Inches and a half. The aperture of the Eye-lids, three quarters of an Inch. From one corner of the Mouth, to the other, Two Inches and a quarter. From the middle of the upper Lip to the Eye-brow, 'twas two Inches three quarters. From the Eye-brow to the *Occiput* Seven Inches and a half. The Perpendicular Diameter of the Ear from the Top to the Lobe, was Two Inches and a half. The Horizontal Diameter of the Ear was an Inch and half. The Verge or Compafs of the Ear about, was near Five Inches and a half. Where the Ear was faftened to the Head, it meafured above an Inch and half. From the *Clavicula* or Collar-Bone, to the *Penis,* Ten Inches. From the *Cartilago Enfiformis* to the Navel, Three Inches and a half. From the Navel to the *Penis,* Three Inches. The diftance between the two Teats, Three Inches

---

(28) Albertus De *Animal.* lib. 22. p. 224. (29) Galen. *de Anat. Adminift.* lib. 1. cap. 2. p. m. 26.

and

and a quarter. The length of the Arm, from the Shoulder to the end
of the Fingers, Seventeen Inches. The Girth of the Shoulder about
the middle, Four Inches and a quarter ; of the Arm near the Elbow,
Five Inches. The Hand from the Wriſt to the end of the Middle Finger, meaſured Five Inches and an half. The Thumb was an Inch and a
quarter long ; the Fore-Finger Two Inches, the Middle-Finger Two
Inches and an half ; the Ring-Finger Two Inches and a quarter, and
the Little Finger One Inch and an half long. The Girth of the Thumb
and the Little Finger, was One Inch ; the Girth of the other Fingers
was an Inch and a quarter. The Palm of the Hand was Three Inches
long, and an Inch and three quarters broad.

From the head of the Thigh-Bone to the Heel, it meaſured Twelve
Inches : From the Heel to the end of the Middle-Toe (which was the
longeſt) Five Inches three quarters. The Girth of the Thigh was Six
Inches ; of the Leg at the Calf, Four Inches and a quarter ; of the Foot
at the ſetting on of the Great Toe, near Five Inches. The Great Toe
was an Inch and half long, the Fore-Toe One Inch, the Middle-Toe
an Inch and half, the Third Toe an Inch and a quarter, the Little Toe
One Inch long. The Sole of the Foot, about the ſetting on of the
Great Toe (where 'twas broadeſt) was Two Inches over ; but nearer
the Heel, 'twas an Inch and half broad. The Girth of the Great Toe,
where biggeſt, an Inch and half ; the other Toes were an Inch about.
Theſe Meaſures were taken before the Skin was ſtrip't off, in the *Skeleton*,
or the Skin ſtuff'd, they may prove otherwiſe.

And having now given theſe *Dimenſions* of the whole, and of moſt
of the External Parts ; you will the better conceive the exact ſhape of
this wonderful *Animal* by the *Figures* I have cauſed to be made of it.
As the *Firſt Figure* repreſents our *Pygmie* erect, where you have a *view*
of all the *Fore-Parts.* Being weak, the better to ſupport him, I have
given him a Stick in his Right-Hand. But our *Figure* being made after
he was dead, the *Head* ſeems too much fallen in between the Shoulders,
as if it had a very ſhort or little or no *Neck*, which takes off from the
Beauty of the *Figure* ; but this is rectified and mended in the Figure of
the *Skeleton*, where you will ſee the *Neck* proportionate. The *Head*
here is large and globous ; the *Ears* ſtanding off, not lying cloſe. The
Face looks like an Old wither'd Man's, which without doubt was render'd much more ſo, by an *Ulcer* it had in one of it's *Cheeks*, occaſioned
by a Fall it had on Ship-board upon a Cannon, which forced out one
of it's Teeth ; and the *Jaw-bone* afterwards proving carious, it might
haſten it's Death. The riſing of the *Cranium* juſt under the *Eye-lids*, as
I have remark'd, is different from what is in a Man, and renders the *Face*
harder ; as does likewiſe it's flat *Noſe*, and the *Upper Jaw* being more
prominent, and leſſer ſpread, than in a *Man* ; and it's *Chin* or *Under Jaw*
being ſhorter. The *Eyes* were a little ſunk, the *Mouth* large, the *Teeth*
perfectly Humane. The *Face* was without Hair, and the Colour a little
tawny ; the *Skin* on the reſt of the Body was white.

The

The *Shoulders* are fpread and large, the *Thorax* or *Breaft* extended altogether like a *Man's*, the *Mammæ* and *Teats* the fame; the *Belly* was lank and pinch'd in, not prominent, by reafon of it's illnefs; but here it held a more proportionable breadth to a *Man's*, than a *Quadruped's*. The *Arms* were longer than in a *Man*, and fo were the *Palms* of the Hands; but the *Thumb* was much lefs, the *Nails* exactly like a *Man's*, and the *Navel* the fame. The *Penis* was different, as we fhall hereafter fhew. Here was no *Scrotum*, but the *Teftes* were contained in the Region of the *Pubis* under the Skin, which made it here more protuberant. The *Thighs* and *Legs* were fomewhat divaricated or ftradling, for want of ftrength, either from it's illnefs, or being but young. We obferved *Calves* in it's Legs; the *Feet* long, as likewife the *Toes*, which were liker Fingers; and the *Great Toe* exactly like a *Thumb*, more than that on the *Hand*.

The *Second Figure* reprefents the *hinder Parts* of this Creature, in an Erect Pofture likewife. Where may be obferved, the Globous Figure and largenefs of the *Head*, with the *Ears* ftanding off; the curious fhape and ftraitnefs of the *Back*, and how it fpreads. At the *Os Coxygis* there is a little Protuberance, but nothing like a *Tail*.

In this *Figure* I have reprefented him with the *Fingers* of one Hand *bended*, as if kneeling upon his Knuckles, to fhew the Action, when he goes on all four: For the Palms of his Hands never touch the Ground, but when he *walks* as a *Quadruped*, 'tis only upon his *Knuckles*. The other Hand is holding a Rope, to fhew his Climbing; for he will nimbly run up the Tackle of a Ship, or climb a Tree: And having this hold, he is the better fupported, to fhew the *Sole* of the left Foot, and the *Heel* there; on account of which Heel it may be thought a *Foot*: But the *Great Toe* being fet off fo far from the range of the others, and they all being fo large and long, it more refembles a *Hand*, as has been obferved.

If we compare *our Figures* with thofe given by *Tulpius*, *Bontius*, and *Gefner*, we fhall find a great difference. That of *Tulpius* feems the moft Natural; but being made fitting, it does not fo well reprefent the Proportions of the feveral Parts. The *Chaps* or *Roftrum* is longer, and 'tis lefs hairy in the fore-parts than ours. The *Mammæ* are larger and pendulous, and the *Belly* more protuberant. *Dapper*, (30) in his Defcription of *Africa*, has borrowed this *Figure* from *Tulpius*, without naming him, as likewife his *Defcription*, which is the fame. For avoiding the often quoting it, I will here Tranfcribe *Tulpius's* Account: But why I think it not a *Satyr*, as *he* and *Dapper* make it, I will give my Reafons in the *following Effay*. *Tulpius* his words are thefe: (31)

---

(30) Dapper *Defcript. de l' Afriqu.* p. m. 365.　(31) *Obfervat. Med.* lib. 3. cap. 56.

D　　　　　　　　　　　　　　　　　*Quamvis*

Quamvis extra forum Medicum, attexam tamen huic tela, Satyrum Indicum; noftrâ memoriâ, ex Angolâ delatum : & Frederico Henrico, Arau-fionenfium Principi, dono datum. Erat autem hic Satyrus quadrupes : fed ab humanâ fpecie, quam præ fe fert, vocatur Indis Orang-Outang : five homo Sylveftris, uti Africanis Quoias morrou. Exprimens longitudine puerum tri-mum; ut craffitie fexennem.

Corpore erat nec obefo, nec gracili, fed quadrato : habiliffimo tamen, ac perniciffimo. Artubus verò tam ftrictis, & mufculis adeò vaftis : ut quidvis & anderet, & poffet. Anteriùs undique glaber : at ponè hirfutus, ac nigris crinibus obfitus. Facies mentiebatur hominem : fed nares fimæ, & aduncæ, rugofam, & edentulam anum.

Aures verò nihil difcrepare, ab humanâ formâ. Uti neque pectus; or-natum utrinque mammâ prætumidâ (erat enim fexûs fœminini); venter habebat umbilicum profundiorem; & artus, cum fuperiores, tum inferiores, tam exactam cum homine fimilitudinem : ut vix ovum ovo videris fimilius.

Nec cubito defuit requifita commiffura : nec manibus digitorum ordo : ne-dum pollici figura humana : vel cruribus furæ : vel pedi calcis fulcrum. Quæ concinna, ac deceus membrorum forma, in caufsâ fuit, quod multoties incede-ret erectus : neque attolleret minùs gravatè, quàm transferret facilè, quale-cunque, graviffimi oneris, pondus.

Bibiturus prehendebat canthari anfam, manu alterâ; alteram verò vafis fundo fupponens, abftergebat deinde madorem labiis relictum, non minùs adpofitè, ac fi delicatiffimum vidiffes aulicum. Quam eandem dexteritatem obfervabat utique cubitum iturus : Inclinans quippe caput in pulvinar, & corpus ftragulis convenienter operiens, velabat fe haud alitèr, ac fi vel mol-liffimus illic decubuiffet homo.

Quin imò narravit aliquandò affini noftro, Samueli Blomartio, Rex Sam-bacenfis, Satyros hofce, præfertim mares, in Infulâ Bornæo, tantam habere animi confidentiam, & tam validam mufculorum compagem : ut non femel impetum fecerint, in viros armatos : nedùm in imbellem, fœminarum, puella-rumve, fexum.

Quarum interdùm tàm ardenti flagrant defiderio : ut raptas non femel con-ftuprarint. Summè quippe in venerem funt procliues (quod ipfis, cum li-bidinofis veterum Satyris commune) imò interdùm adeò protervi, ac falaces : ut mulieres Indicæ, proptereà vitent, cane pejus & angue, faltus, ac luftra, in quibus delitefcunt impudica hæc animalia.

Dapper, who hath tranfcribed this Account of *Tulpius*, (as I faid) but without taking any notice of him, makes this Preface to it. " *The* " Quoias Morrou (*of which I have fpoken in the Kingdom of* Quoia) " *are bred likewife in the Kingdom of* Angola. *This Animal, as it hath a* " *great deal of a Man, fo a great many have thought it to be the Iffue of a* " *Man and an Ape : But the* Blacks *themfelves reject this Opinion.* Now in the Place that *Dapper* refers to, he feems to give it as the Opinion of the *Blacks*, that they are the Iffue of Men; but that by their always

living

living in the Woods, they are become *half-Beafts.* I fhall tranfcribe his Words, and fo have done with him: (32) *On trouve dans les bois une Efpece de Satyre que les Negroes appellent Quoias-Morrou, & les Portugais, Salvage. Ils ont la tête groffe, le Corps gros et pefant, les bras nerveux, ils n'ont point de quevë, et Marchent tantôt tout droit, et tantôt à quatre pieds. Les Animaux fe nourriffent de fruits et de Miel Sauvage, & fe battent à tout moment les uns contre les autres. Ils font iffu des Hommes, à ce difent les Negroes, mais ils font devenus ainfi demi-bêtes en fe tenant toûjours dans les Forêts. On dit qu'ils forcent les femmes & les filles, & qu'ils ont le courage d' attaquer des Hommes armez.*

We will now examine *Jacobus Bontius's Figure,* and compare it with ours: And tho' he tells us, that he had feen fome of both Sexes that went erect, efpecially that *Female one,* whofe *Effigies* he here gives us; yet I can't but think, he indulged more his Fancy herein, than copied the true Life; or at leaft it was much different from ours. For ours had no fuch long *Hair* on the Head, and all round the Face; the *Face* of our *Pygmie* was not fo flat and round, nor the *Nofe* and *Under-Lip* fo rifing: The large *Breafts* in his, anfwers the Defcription which is given of it by others; ours being a *Male,* had but fmall ones. But the *Armes* in our *Pygmie* (as 'tis in the *Ape-kind*) were much longer than they are reprefented in his *Figure,* and the *Feet* are altogether different; for he makes them exactly like *Humane Feet,* and nothing like *Hands,* which is fo Remarkable a thing in all thefe *Animals,* that this Miftake of it felf, is enough to difcountenance the Truth of his *Picture,* and render it fufpected. I fhall not take notice, how ill the *Hair* is drawn, nor make any further Remarks upon the ftructure of the *Limbs,* fince I confefs I do miftruft the whole *Reprefentation.* But becaufe he hath ex- prefs'd, that this Creature had fo much Modefty, I have added to his *Figure* what becomes that Character.

That Figure in *Conradus Gefner,* (33) which he tells us he had out of a *German* Book, wrote about the Holy Land, in fome Particulars I think more exact and juft: For here he makes the *Feet* like *Hands,* the *Legs* more divaricated, the *Face* longer, and the *Roftrum* more extended. But the *Arms* are too fhort, and I do not know for what reafon there is a *Tail* clap't on, which fits untowardly enough, which muft be furely an Addition of the *Painter*; or if there is any fuch Creature in Nature, it muft be of another Family, different from ours.

However, I have caufed all thefe *Figures* to be copyed, that they may be the eafier compared : But fince they are fo difagreeing, as are likewife

(32) Dapper *ibid.* p. m. 257. (33) *Hift. de Quadruped.* p. m. 859.

the

the Defcriptions they give of them, it fufficiently juftifies my Complaint
of the uncertainty we have of the true *Animal*, that they are difcourfing
about ; fince the fame Name probably may be given to different *Species*
of the *Ape*-kind. Now *Orang-Outang*, or *Homo Sylveftris*, or the *Wild
Man*, being a *General Name*, I have given it alfo to our Subject : Tho' I
confefs I am not fully fatisfied whether it be exactly the fame with that
of *Tulpius* or *Bontius* ; or even whether that of *Bontius* be the fame
with that of *Tulpius*. For *Bontius* his Account is fo very imperfect,
that from thence one cannot make a fafe Conclufion ; and I rather
fufpect the contrary : For *Bontius* defcribes it with foft, tender Paffions ;
Tulpius and *Dapper* make it Warlike and Fighting. *Bontius*'s words are
thefe : (34) *Aft quod majorem meretur admirationem, vidi ego aliquot utri-
ufque fexus erecte incedentes, imprimis eam (cujus Effigiem hic exhibeo)
Satyram femellam, tanta verecundia ab ignotis fibi hominibus occulentem,
tum quoque faciem manibus (liceat ità dicere) tegentem ubertimque lachry-
mantem, gemitus cientem, & cæteros humanos actus exprimentem, ut nihil
ei humani deeffe diceres, præter loquelam. Loqui verò eos eafque poffe, Javani aiunt, fed non velle, nè ad labores cogerentur : ridicule mehercules.
Nomen ei indunt Ourang Outang, quod Hominem Sylvæ fignificat, eofque
nafci affirmant è Libidine Mulierum Indarum, quæ fe Simiis & Cercopithe-
cis deteftandâ libidine mifcent :*

Nec pueri credunt, nifi qui nondum ære lavantur.

And then adds, that in *Borneo* there are thefe *Wild Men*, and with
Tails, but much fhorter than that pictured in *Gefner*. *Porrò in Infulâ
Bornco* (faith *Bontius*) *in Regno Succodana dicto, à noftris Mercatoribus
propter Oryzam & Adamantes frequentato, Homines montani Caudati in interioribus Regni inveniuntur, quos multi è noftris in Aulâ Regis Succodanæ
viderent. Cauda autem illis eft prominentia quædam offis Coccygos, ad quatuor, aut paulò ampliùs, digitos excrefcens, eodem modo, quo truncata cauda
(quos nos* Spligiones *vocamus*) *fed depilis.*

'Tis for this Reafon therefore, that I might more particularly diftinguifh our *Animal*, that I have call'd it a *Pygmie* ; a *Name* that was formerly given to a fort of *Ape*, as I fhall prove. But the *Poets* and *Hiftorians* too of former Ages have invented fo many improbable Stories
about them, that they have rendred the whole *Hiftory* concerning them
ridiculous, and not to be believed. We fhall therefore endeavour to
diftinguifh the *Truth* from the *Fables* in the following *Effay*.

The *Baris* or *Barris* likewife feems to be an *Ourang Outang*, or a *Wild
Man* ; but whether exactly the fame with ours, I will not determine,
but leave to farther Enquiry. For all the Accounts concerning it that I

---

(34) Jac. Bontji, *Hift. Nat. & Med.* lib. 5. cap. 32.

have

have at present met with, relate rather it's Docility and Actions, and the Servile Offices 'tis capable of performing in a Family, than any thing particular as to the Description of the Body; only in general that 'tis an Ape like a Man. Thus *Peter Gaffendus* (35) in the *Life* of *Peiresky* tells us, that in *Java Major* were observed by the *Sieur de Saint-Amant*, *Animalia quæ fovent Naturæ homines inter & simias media*; which being doubted of, *Peiresky* produced a Letter from *Natalis* or *Noël*, a Physician who lived in *Africa*, which gave him this Account. *Effe in Guineâ Simias, barbâ procerâ, caudaque, & pexâ propemodùm venerabileis, incedere ipsos lentè, ac videri sibi præ cæteris sapere: qui maximi sunt, & Barris dicuntur, pollere maximè judicio; semel duntaxat quidpiam docendos; veste indutos illicò bipedes incedere; scitè ludere fistulâ, Citharâ, aliisque id genus (nam quod everrant domum, convertant veru, pinsant in mortario, aliaque ratione famulatum præstant, haud reputari admodum) fæminas denique in iis pati menstrua, & mares mulierum esse appetentissimos.* He likewise produced other Letters from *Arcosius* or *d'Arcos*, which related what happened to one of *Ferraria* when he was at *Angola*, the Country from whence our *Animal*, as likewise that of *Tulpius* came. I will give it in *Gaffendus*'s words: *Incidit nempè quâdam die in Nigritam Canibus venantem Homines ut visum, Sylvestreis. Capto, cæsoque illorum uno, inhumanitatem Nigritæ increpuit, qui in suum genus ità sæviret. Ille verò, falleris, inquit, nam hic non est homo, sed bellua homini persimilis. Quippe solâ pascitur herbâ, intestinaque Ovina habet, quod ut credas meliùs, rem ecce; simulque abdomen aperuit. Sequenti die rursus venatum, captique mas & fæmina: huic Mammæ ad pedis longitudinem: cætera mulieri simillima fuit; nisi quod Intestina quoque herbis oppleta, & cujusmodi Ovis, habuit. Totum utique pilosum Corpus, sed pilo brevi, ac satis leni.*

Our *Animal* was not so bearded, as that of *Natalis*; and what *Arcosius* relates of his *Wild Man*, or *Barris*; as it's feeding upon Grass, and having it's *Intestines* like a *Sheep*'s, all this is far different from ours; tho' as to it's docility and capacity of performing those Actions mentioned, I can't but think our Subject might easily have been taught to do them; and, it may be, others too of the *Ape*-kind, tho' different: As there are wonderful Instances of this kind given of them by *Nierembergius* (36) and others. *Dapper*'s (37) Description is much the same. There is a sort of Ape (saith he) call'd Baris, which they take when young, and breed them up, and make them so tame, that they will do almost all the Work of a Slave: For they go ordinarily upright as Men do; they will beat Rice in a Mortar, carry Water in a Pitcher, and shew such pretty Actions of Address, that they extreamly divert their Masters. And in *Nierembergius* (38) there is much the same Account. " In Guineâ scribit P. Jarricus existere Simias, quæ instar famuli in Pilâ tundant quæcunque in

(35) *Lib.* 5. p. m. 171. (36) *Hist. Nat.* lib. 9. cap. 44. (37) *Dapper Descript. de l'Afrique.* p. 249. (38) *Hist. Nat.* l. 9. cap. 45.

" eam

" eam imponuntur, quæ aquam à fluviis in Hydriis capite domùm defe-
" rant, ità tamen ut ubi primùm domûs fores attigerint, illicò Hydriis
" exonerandæ fint, alioqui eas excidere, cafuque ifto frangi, atque tùm
'· clamoribus ac fletu compleri univerfa. Neque ifta modo, fed plurima
' item alia obire de domefticis minifteriis dicuntur hi Simij Baris. To-
" rofi funt & robufti.

But all this does not fufficiently inform us of the particular fhape,
ftructure, and make of the Body and the feveral Parts of this *Animal*,
fo as to be fully certain whether it be the fame, or a different Creature
from the *Ourang-Outang.* And tho' I have mentioned it, as a Conjecture that probably the *Baris* might be, what we call a *Drill*, yet I
own it as an uncertainty, fince I have not met with what can juftifie, or
fully fatisfie me herein.

The *Pongo* likewife which is defcribed by *Purchas*, as a fort of *Wild
Man*, is different from our Subject; as it may be alfo from the reft hitherto mentioned. The Reafon, therefore, why I infert the Defcription of this, as likewife of the others, I own to be, that hereby I might
excite fome *Inquifitive Obfervers* to give us a truer Account of this large
and noble *Species* of *Animals.* 'Tis an Enquiry that would recompence
their Curiofity with abundance of Satisfaction, by the many and ufeful
Difcoveries that they would make, and extreamly enrich the *Natural
Hiftory* of *Animals*, whofe enlargement, I think, in this *Inquifitive Age*,
hath not advanced fo much as that of *Botanie.* For how great Diligence
hath been ufed of late, to ranfack both the *Indies*, to pry into all the
Corners of the World, both inhabited, and uninhabited, to find out a
new *Plant*, not before defcribed ? And with what great Expence, and
how magnificently are their *Figures* Printed ? And how little hath been
done in the Improvement of the *Hiftory* of *Animals* ? Not that I any
ways diflike the former, but the latter being a *Nobler* Subject, I can't but
recommend it, as deferving the Labours of the *Curious* likewife ; and if
any, this kind, I think, which comes fo near to a Man, may befpeak
the preference.

But I beg the Reader's Pardon for this Digreffion. *Purchas*'s (39) words
are thefe : *This* Pongo *is in all Proportions like a Man, but that he is more
like a Giant-Creature, than a Man : For he is very tall, and hath a Man's
Face, hollow-eyed, with long Hair upon his brows. His Face and Ears are
without Hair, and his Hands alfo. His Body is full of Hair, but not very
thick, and it is of a dunnifh colour. He differeth not from a Man, but in
his Legs, for he hath no Calf. He goeth always on his Legs, and carries his
Hands clafped on the Nape of his Neck, when he goeth upon the Ground.*

---

(39) Purchas *Pilgrims*, Part. 2. l. 7. cap. 3. §. 7.

They *fleep in the Trees, and build fhelters for the Rain. They feed upon Fruits that they find in the Woods, and upon Nuts ; for they eat no kind of Flefh. They cannot fpeak , and have no Underftanding, no more than a Beaft. The People of the Country, when they Travel in the Woods, make Fires, where they fleep in the Night : And in the Morning when they are gone, the Pongoes will come and fit about the Fire, till it goeth out ; for they have no Underftanding to lay the Wood together. They go many toge-ther, and kill many* Negroes *that Travel in the Woods. Many times they fall upon* Elephants, *which come to feed where they be, and fo beat them with their clubbed Fifts, and pieces of Wood, that they will run away roaring from them. Thefe* Pongoes *are never taken alive , becaufe they are fo ftrong , that Ten Men cannot hold one of them : But yet they take many of their Young Ones with poifoned Arrows. The Young* Pongo *hangeth on his Mo-ther's Belly, with his Hands faft clafped about her ; fo that when any of the Country People kill any of the Females, they take the Young one which hangeth faft upon his Mother. When they die among themfelves, they cover the Dead with great heaps of Boughs and Wood, which is commonly found in the Forrefts.*

Our *Pygmie* had *Calves* in his Legs, tho' not large, being emaciated ; and it being young, I am uncertain to what height in time it might have grown ; tho' I cannot think to the juft Stature (if there be any fuch) of a Man. For different Nations extreamly vary herein, and even thofe of the fame. Nor did our *Pygmie* feem fo dull a Creature as thefe *Pongoes*, but on the contrary, very apprehenfive, tho' nothing fo robuft and ftrong as they are reprefented to be.

I fhall only further add what *le Compte*, a Modern Writer, tells us of the *Savage Man*, and fo I think I fhall have done : For this Argument . is fo Fruitful, that one does not know when to conclude. (40) *Lewis le Compte* therefore in his *Memoirs and Obfervations upon China*, tells us, *That what is to be feen in the Ifle of* Borneo, *is yet more Remarkable, and furpaffeth all that ever the Hiftory of Animals hath hitherto related to be the moft admirable, the People of the Country affure us, as a thing notorioufly known to be true : That they find in the Woods a fort of Beaft, called the* Savage Man ; *whofe Shape, Stature, Countenance, Arms, Legs, and other Members of the Body, are fo like ours, that excepting the Voice only, one fhould have much ado not to reckon them equally Men with certain Barbarians in* Africa, *who do not much differ from Beafts.*

This Wild or Savage Man, *of whom I fpeak, is endued with extraordi-nary ftrength, and notwithftanding he walks but upon two Legs ; yet is he fo fwift of Foot, that they have much ado to out-run him.* People of Quality

(40) Pag. m. 510.

Courſe him, as we do Stags here, and this ſort of Hunting is the King's
uſual Divertiſement. His Skin is all hairy, his Eyes ſunk in his Head, a
ſtern Countenance, tanned Face ; but all his Lineaments are pretty proportionable, although harſh, and thickned by the Sun. I learn'd all theſe Particulars from one of our French Merchants, who hath remained ſome time
upon the Iſland. Nevertheleſs, I do not believe a Man ought to give much
Credit to ſuch ſort of Relations, neither muſt we altogether rejeēt them as
fabulous ; but wait till the unanimous Teſtimonies of ſeveral Travellers may
more particularly acquaint us with the Truth of it.

Paſſing upon a time from China to the Coaſt of Coramandel, I did my
ſelf ſee in the Straits of Molucca a kind of Ape, that might make pretty
credible that which I juſt now related concerning the Savage Man.

It marches naturally upon it's two hind Feet, which it bends a little, like
a Dog's, that hath been taught to Dance, it makes uſe of it's two Arms as we
do ; it's Viſage is in a manner as well favoured, as theirs of the Cape of Good
Hope ; but the Body is all covered with a white, black, or grey Wool : As
to the reſt, it cries exaētly like a Child ; the whole outward Aētion is ſo Humane, and the Paſſions ſo lively and ſignificant, that dumb Men can ſcarce
expreſs better their Conceptions and Appetites. They do eſpecially appear to
be of a very kind Nature ; and to ſhew their Affeētions to Perſons they know
and love, they embrace them, and kiſs them with tranſports that ſurpriſe a
Man. They have alſo a certain motion, that we meet not with in any Beaſt,
very proper to Children, that is, to make a noiſe with their Feet, for joy or
ſpight, when one gives, or refuſes them what they paſſionately long for.

Although they be very big, (for that I ſaw was at leaſt four Foot high)
their nimbleneſs and ſlight is incredible ; it is Pleaſure beyond expreſſion to
ſee them run up the Tackling of a Ship, where they ſometimes play, as if they
had a particular knack of Vaulting to themſelves, or as if they had been
paid, like our Rope-Dancers, to divert the Company.

Sometimes ſuſpended by one Arm, they poiſe themſelves for ſome time neg-
ligently to try themſelves, and then turn, all on the ſudden, round about a
Rope, with as much quickneſs as a Wheel, or a Sling that is once put in motion ; ſometimes holding the Rope ſucceſſively with their long Fingers, and
letting their whole Body fall into the Air, they run full ſpeed from one to the
other, and come back again with the ſame ſwiftneſs. There is no Poſture
but they imitate, nor Motion but they perform ; bending themſelves like a
Bow, rowling like a Bowl, hanging by the Hands, Feet, and Teeth, according to the different Fancies which their whimſical Imagination ſupplies them
with, which they Aēt in the moſt diverting manner imaginable ; but their
Agility to fling themſelves from one Rope to another, at Thirty and Fifty Foot
diſtance, is yet more ſurpriſing.

In

In this Character there are several things I could take notice of, and I may hereafter have occasion to refer to some of the Particulars; But what is mention'd of it's *Cry*, like a Child's ; and it's expressing the *Passions* of Joy and Grief, by making a Noise with it's Feet, is agreeable enough to the Relation I had of our *Pygmie* : For I heard it *Cry* my self like a *Child* ; and he hath been often seen to kick with his Feet, as Children do, when either he was pleased or angered.

We shall now proceed to the *Anatomy*, which in a *History* of *Animals*, is certainly the most Necessary, most Significant, and Instructive Part. Nor can I see, how an *History* of *Animals* can be well wrote without giving the *Dissection* of the *Inward Parts* : 'Tis as if one should undertake to describe a *Watch*, and at the same time, take notice only of the *Case* or Cover, and tell what fine Garniture there is about it ; but inform us nothing of the admirable Contrivances of the *Wheels* and *Springs within*, which gives it Life and Motion. *Galen* (41) thought the *Dissection* of *Apes* very useful for the understanding the Structure of the *Parts* in *Humane Bodies* ; and recommends it to his Scholars to Practice themselves herein. Not that he only dissected *Apes*, (as *Vesalius* oftentimes charges him with) or preferred it before the Dissection of *Humane Body :* But where that could not be had, he advises them to get *Apes*, and dissect them ; especially those that come nearest to a *Man*. Had he known our *Pygmie*, no doubt but he would have preferred it , for this purpose, as much beyond the *Ape*, as he does the *Ape* beyond the *Cynocephalus*, and all other *Animals*. For, as we shall observe, there is no *Animal*, I have hitherto met with, or heard of, that so exactly resembles a *Man*, in the Structure of the *Inward Parts*, as our *Pygmie* : But where it differs, (as I have remark'd) there it resembles an *Ape* ; being different both from a *Man* and an *Ape* : And in many things agreeing with both of them.

The *Skin* of the whole Body of our *Pygmie* was whitish ; but that on the *Head* was tawny, and of a darker colour. 'Twas thin, but strong, and adhered pretty firmly, and more than usually to the Flesh ; it's greatest adhæsion was at the *Linea alba*, and in the *Palms* of the Hands, and the *Soles* of the Feet, and in the *Fingers* and *Toes* ; as it is in *Men*. In the *Skin* of the *Arm-pits*, I observed those *Glandulæ Cutaneæ Axillares*, which secrete that *Orange*-coloured *Liquor*, which in some Men stains the Shift here, with that colour. I call them *Cutaneæ*, to distinguish them from those larger *Glands*,that lie bedded under in the Fat, and are call'd *Glandulæ Axillares*. For these I have observed to be *Lymphatic Glands* ; and have traced the *Lymphaducts* thence to the head of the *Ductus Thoracicus*, where they empty themselves.

(41) *De Anat. Administr. lib.* 1. *cap.* 2. p. m. 27.

E

Together with the *Skin*, we took off the *Mammæ* or *Breaſts*, which ſtuck cloſe to it : And in our Subject, being a *Male*, they were but ſmall and thin ; yet I could plainly perceive they were made up of abundance of ſmall *Glands*. I have already mentioned, how large the *Breaſts* are in the *Female Orang-Outang*, and the *Baris*, ſo that no *Woman's* are larger. As to their *Situation*, and their being placed upon the *Pectoral Muſcles*, this I find is common to the *Ape-kind* : And they are ſo deſcribed by the *Pariſians* (42) in the *Monkeys* they diſſected ; as alſo in the *Apes* diſſected by *Drelincourt* : (43) And becauſe I ſhall have frequent occaſion of re-ferring to theſe Authors, unleſs I ſignifie otherwiſe, I ſhall always mean the Places here quoted, without mentioning them any more.

In *Brutes*, next under the *Skin*, lies a *Muſculous Membrane*, which therefore is call'd *Panniculus Carnoſus*, which gives a motion to it, where-by they can ſhove off what offends them. In *Man* 'tis otherwiſe ; for next to the *Skin*, lies the *Membrana Adipoſa* ; or the *Fat*, and under that, the *Membrana Carnoſa* : And the ſame I obſerved in our *Pygmie* ; for the *Fat* here lay next to the *Skin*. *Drelincourt*, in the *Apes* he diſſected, obſerved the *Panniculus Carnoſus* next to the *Skin*, as 'tis in *Brutes*. For in the *Male Ape*, he tells us, *Adeps nullus inter Panniculum Carnoſum & Cutim* ; and in the *Female*, *Panniculus Carnoſus cuti cohærens, nullo adipe inter-jecto, Adipoſus nullus*. So that in this Particular, our *Pygmie* is like to a *Man*, and different from the *Ape-kind*.

Having ſeparated the *Skin* and *Membrana Adipoſa*, which in our Sub-ject was not very thick, it being emaciated by it's illneſs, we come now to the *Muſcles*. But I ſhall referve my ſelf to treat of them in the *My-ſlogy*. Next under the *Muſcles* was the *Peritonæum*, a Common Mem-brane, that lines all the inſide of the *Abdomen*, and ſends a common outward Membrane to all the *Viſcera* contained therein, and ſo ſecures their Situation. In this *Membrane* in *Quadrupeds* there is in the Groin of each ſide, a Perforation, or rather a *Proceſſus*, by which the *Seminal Veſſels* paſs down to the *Teſtes* in the *Scrotum*, as is very plain in *Dogs* and other *Animals*. But in *Man*, whoſe Poſture is erect, 'tis otherwiſe. For here theſe *Veſſels* paſs between the two *Coats*, that make up this *Mem-brane*, the *Peritonæum* ; ſo that the *inward* Coat, that reſpects the Cavity of the *Abdomen*, is altogether entire, and continued, and 'tis only the *outward* Coat that is protruded into this *Proceſs* ; and this for a very good Reaſon. For otherwiſe, a *Man*, whoſe Poſture is erect, would be conſtantly liable to an *Hernia*, or a *Rupture* ; which happens when this *inward* Coat is protruded down likewiſe ; and if there be a deſcent of the *Inteſtines*, 'tis then call'd *Entero-cele* : If of the *Omentum*, *Epiploo-cele*. In our *Pygmie* I obſerved the *Peritonæum*, in this reſpect, to be

(42) Memoirs for a Natural Hiſtory of Animals, *p.* 162, *&c. Engliſh* Tranſlation. (43) *Apud* Ger. Blaſii, *Anot. Animal.* cap. 33. pag. 109, *&c.*

formed

formed exactly as 'tis in *Man* ; and to be entire, and not protruded ; as if *Nature* did defign it to go *erect*. In *Apes* and *Monkeys* 'tis otherwife. So *Blafius* (44) obferved in the *Ape* he diffected, *Proceffus Peritonæi* (faith he) *codem modo hic fe habet, ac in Cane. Datur & hic facilis via ftylo ex ventre in Proceffum dictum inferendo*. And the *Parifians* have remarked the fame in the *Monkeys* they diffected, which is a notable difference of our *Pygmie's* from the *Ape-kind*, and an agreement with the *Humane*. Hereafter, whenever I mention *Blafius*, unlefs I fpecifie otherwife, be pleafed to take notice, that I refer to this Quotation.

The *Omentum* or *Caul* in our *Pygmie* was very thin and large, falling over and covering moft parts of the Guts. 'Twas faftened a little to the *Peritonæum* in the Left Side. It had but little *Fat*, and was tinged in many places with a deep Yellow Colour, by the Bladder of the *Gall*, as was likewife part of the *Duodenum*. It had numerous Blood-Veffels, and it's adhæfion to the *Stomach, Colon*, and other Parts, as in *Man*. The Remarks the *Parifians* make upon the *Epiploon* or *Omentum* of the *Monkeys* they diffected, were different from our Subject. For they tell us, That *the* Epiploon *was different from that of a Man, in feveral things*. Firft, *It was not faftened to the* Colon *in fo many places, having no con-nexion with the left part of this Inteftine*. Ours I found was faftened juft as 'tis in Man. Secondly, *It had another Ligature, which is not found in Man, viz. to the Mufcles of the* Abdomen, *by means of the* Peritonæum, *which formed a Ligament ; which we have obferved in the* Hind *of* Canada. Ours adhered to the Left fide : *Drelincourt* obferved it in an *Ape*, to be faftened to the Right Side. Both I believe to be accidental, as I have frequently feen it in *Humane* Bodies. And in one *Patient* I found it fixt to the *Peritonæum* in the *Groin*, which gave him a great deal of Pain and Trouble, efpecially when his Bowels were any thing extended with Wind. Thirdly, *The* Parifians *fay, The Veffels of the* Epiploon, *which in Man proceed only from the Vena* Porta, *did neverthelefs in one of our Subjects come from the* Cava, *having there one of the Branches of the* Hy-pogaftrica, *which was united to the Branches of the* Porta. In our *Animal* thefe Veffels came all from the *Porta*, or rather emptied themfelves into it. But they obferving it only in one Subject, and it being different in all other *Animals*, it muft be accidental. Fourthly, *In fine, the whole* Epiploon *was without Comparifon greater than it generally is in Man ; be-caufe that it did not only cover all the Inteftines, which is rarely feen in* Man, *(whatever* Galen *fays) but it even enveloped them underneath, as it does in feveral other Brutes ; where it is frequently feen, that the* Epiploon *is lar-ger than in Man, efpecially in Animals that do run, and leap with a great deal of Agility : As if it were fo redoubled under the Inteftines to defend them, with the reft of the Bowels, against the rude jolts which thefe Parts do*

---

(44) Ger. Blafij *Mifcellan. Anat. Hominis Brutorumque variorum*, &c. in *Octavo*. p. m. 253.

receive

receive in running. It is true, that the *Membranes of the* Epiploon *were entire and continued, as in* Man, *and not perforated like a Net, as they are in the generality of Brutes.* The *Epiploön* or *Caul* in our *Pygmie* was very large, yet I have ſeen the ſame frequently in *Humane Bodies* ; but when they are diſeaſed, 'tis often leſs, and waſted ; ſo that *Galen*'s Obſervation may be true. But methinks the Reaſon they give, why it ſhould be ſo large in Brutes, may be doubted of ; for it being ſo tender a Part, it would be in danger, upon thoſe violent motions, of being broken, had not *Nature* made it looſe below, and free from any adhæſion ; and it being ſo, it cannot perform the Office they aſſign it. *Drelincourt*'s Account of the *Epiploon,* as he obſerved it in the *Female Ape,* I like better. *Epiploon macrum* (ſaith he) *vaſis turgidis involvens Inteſtina omnia, uſque ad pubem, adhærens Extremo Hypochondrio dextro, quâ parte Colon ſubſtratum jecoris limbis. Idem adhæret ventriculi fundo & Colo, ut in homine.* And in the *Male Ape* he diſſected 'twas tinged yellow, as ours was.

We ſhall proceed now to the *Ductus Alimentalis,* at leaſt thoſe parts of it that are contain'd in the *Abdomen, viz.* the *Stomach* and *Inteſtines* ; which I make to be the true *Characteriſk* of an *Animal,* and a *Proprium quarto modo.* For all *Animals* have theſe Parts ; and all that have them, are *Animals.* The *Senſes,* or ſome of them, are wanting in a great many *Animals,* and in ſome we perceive none but that *Univerſal one, Tactus,* yet here we find a *Ventricle* and *Inteſtines.* By theſe Parts 'tis, that the *Animal* Kingdom is principally diſtinguiſh'd both from the *Vegetable* and *Angelick. Vegetables,* 'tis true, receive conſtantly Nouriſhment, and without it, they periſh and decay ; but 'tis in a far different manner ; 'tis not received into ſuch an *Organick* Body, where the Food is prepared and digeſted, and ſo the *Nutritive* parts thereof diſpenſed afterwards into all the Body, and the reſt ejected, as *Excrementitious* ; this is only to be met with in *Animals,* and in all of them. But yet I find there are intermediate *Species* of *Beings* between *Vegetables* and *Animals,* as the *Zoophyta :* the *Hiſtory* of which I could extreamly deſire might be given us ; and can't but think that regularly in compiling a *Hiſtory* of *Animals,* one ſhould commence from them ; and amongſt theſe, no doubt, but that there are ſeveral degrees of *Perfection,* till we come to what might be properly called an *Animal.* I have had no Opportunity of obſerving any of them, but only one ; wherein I could perceive a ſenſible *Motion* and Contraction of ſome of the Parts, but could not diſtinguiſh any thing like the Structure of any of the Parts in an *Animal,* or the *Organs* that belong to them. An Accident diſappointed me of perfecting my Obſervations, otherwiſe I ſhould have communicated what I had diſcovered. But am ſenſible that there are great *Curioſities* here to be met with, if diligently enquired into ; and that they might be, was the occaſion of this *Digreſſion.*

This

This *Canalis Alimentalis* therefore, or *Inductory Vessel* (as I call it, for the Reasons I have often mentioned in my *Anatomical Lectures* at *Chirurgeon's Hall*) is commonly diſtinguiſhed into three Parts ; The *Gula*, the *Ventricle*, and *Inteſtines* : The two latter do lie in the Cavity of the *Abdomen*, the former, in the *Thorax* and *Neck* ; but being but one continued *Canalis*, I ſhall treat of the whole here.

The *Gula* or *Gullet*, by (45) *Tully* (in that excellent *Anatomical* Lecture he gives us, where he is proving a *Providence*) is call'd *Stomachus*. As 'tis alſo by *Celſus*, (46) ſo likewiſe in *A. Gellius*, (47) and frequently by *Hippocrates*. (48) And *Ariſtotle* (49) and *Galen* (50) expreſly tell us, that that Part between the *Fauces* and the *Ventricle*, which the Antients called *Oeſophagus*, after *Ariſtotle's* time, was wont to be call'd *Stomachus*, tho' now this word is more appropriated to the *Ventricle* it ſelf, which *Tully* in the ſame place calls *Alvus*. So true is that of *Horace*,

> (51) *Ut Sylvæ foliis pronos mutantur in annos*
> *Prima cadunt : ita verborum vetus interit ætas.*

However I ſhall follow *Horace's* Rule, ſince Cuſtom now hath appropriated the word *Stomach*, to the *Ventricle*, eſpecially our *Engliſh Tongue*, I ſhall do ſo too.

> *Multa renaſcentur, quæ jam cecidere : cadentque*
> *Quæ nunc ſunt in honore vocabula : ſi volet uſus :*
> *Quem penes arbitrium eſt, & vis & norma loquendi.* (52)

This *Gula* or *Gullet* is a *Hollow Muſcle*, and fitly enough compared to a *Funnel* ; where the *Mouth*, which may be thought a Part belonging to it, being more capacious, firſt receives the *Food*, and prepares it, by chewing, and then forces it down into this Stem or *Pipe*, to convey it to the *Ventricle*. I did not obſerve, upon the Diſſection, any difference of this Part in our *Pygmie*, from that of a *Man*. For as in a *Man*, (and ſo conformable too in other Circumſtances) it paſſed under the *lower Muſcle* of the *Diaphragm*, which by that ſlant running of it's fleſhy Fibres over it, may perform to it the Office of a *Valve*, and prevent the Regurgitation of the Food that way. Which may be the more neceſſary in our Subject, becauſe being uſed to climb Trees, and in coming down, to be *prono Capite*, it might be the more liable to this Accident. But for the better preventing this, I find here, that the Paſſage of the *Gula*, a little above where it empties it ſelf into the *Ventricle*, was ſtraiter, and

---

(45) M. T. Cicero *de Nat. Deorum*, lib. 2. §. 54. p. m. 427. (46) Cornel. Celſus, *de re Med.* lib. 4. cap. 1. (47) A. Gellij. *Noct. Attic.* lib. 17. cap. 11. (48) *Vid. Anat. Fæſii Oeconom. Hipp. in verbo.* (49) Ariſt. *Hiſt. Anim.* lib. 1. cap. 10. §. 108. p. m. 89. & *paſſim alibi.* (50) Galen *de locis affectiis.* lib. 5. cap. 5. p. m. 490. (51) Horace *de Arte Poetic.* verſ. 60. (52) Horace. *Ibid.* v. 70.

the inward Membrane here more rugous than in a Man ; fo that it feemed fomewhat Analogous to a *Valve*. *Drelincourt* defcribes it, in the *Female Ape* he diffeᴄted, thus. *Orificium ejus fuperius, nullâ Valvulâ claufum ; fed interceptum duplici portione Diaphragmatis carnosâ, ab ejus tendinibus oriundâ.*

The *Ventricle* or *Stomach*, as we fhall call this Part, in our *Pygmie*, as to it's Situation and Figure, exaᴄtly reprefented a *Humane Stomach*. When inflated, from the entrance of the *Gula* along the upper part to the *Pylorus*, it meafured Two Inches and three quarters. Meafuring with a Thread from the *Pylorus* along under the *Fundus*, up again to the entrance of the *Gula*, I found it to be Fifteen Inches ; in all , near Eighteen Inches. The length of the *Stomach* in a ftrait Line, was Six Inches and an half ; and it's breadth in a ftrait Line, where broadeft, near Four Inches. The Girth of the *Stomach* in the middle, was near Twelve Inches. So that I found the *Stomach* large, in Proportion to the bulk of the Body. It had numerous *Blood-Veffels*, fpreading themfelves all over, as in a *Man's* ; and I could plainly perceive the Inofculations of large Trunks of the *Coronary* Branches, with thofe that defcended from the upper Parts.

The *Parifians* obferved in their *Monkeys*, That the *Ventricle* did likewife differ from a *Man's*, it's *inferiour Orifice being very large and low ; for it was not elevated fo high as the fuperiour, as it is in a Man*. I did not obferve this in our *Pygmie*. So *Drelincourt* tells us in the *Female Ape*, *Ventriculus rugis interius nullis gaudet* ; and fome other Particulars he takes notice of. But there was nothing in ours, that I obferved, different from a *Man's*.

As to their *Food*, I find it very different in the *Ape-kind* ; as in part appears by what I have already mentioned of the *Ourang Outang*, the *Baris*, the *Pongo*, &c. So that I can't but think, (like a Man) that they are *omnivorous*. What chiefly our *Pygmie* affeᴄted, when *Wild*, I was not informed of ; after it was taken, and made tame, it would readily eat any thing that was brought to the Table ; and very orderly bring it's Plate thither, to receive what they would give him. Once it was made Drunk with *Punch*, (and they are fond enough of ftrong Liquors) But it was obferved, that after that time, it would never drink above one Cup, and refufed the offer of more than what he found agreed with him. Thus we fee *Inftinᴄt* of *Nature* teaches Brutes *Temperance* ; and *Intemperance* is a Crime not only againft the *Laws* of *Morality*, but of *Nature* too.

Jacobus

Jacobus Bontius (53) tells us, that the *Bezoar-ftone* is bred in the Stomachs of *Apes*, as well as *Goats*, and he prefers it as the beft. *Porrò vidi* (faith he) *& Lapides* Pa-zahar *natos in ventriculis Simiorum, qui teretes funt & longitudinem digiti aliquandò excedunt, qui præftantiffimi omnium cenfentur.* Pa-zahar, he tells us a little before, fignifies in the *Perfian, contra venenum,* whence may come the word *Bezoar.* *Joh. Georg. Volchamerus* (45) takes notice of one he had from *Grimxius* out of the *Baboon*-kind, as big as a Walnut. And in the *Scholium* on that Obfervation, *Joh. Bapt.* Tavernier's (55) Travels are quoted, where he prefers two Grains of this, before fix of the *Goat-Bezoar.* Tho' *Philip. Baldæus,* in his Defcription of *Malabar* and *Cormandel,* does efteem it much cheaper. *Cafper Bauhinus* hath wrote a diftinct Treatife of the *Bezoar-ftone,* to whom I refer my Reader; and fhall only farther obferve of it, that I think this *Medicine* ought not to be defpifed, becaufe in Health a Man may take a large Quantity of it, without any Injury; for I have evidently feen in the greateft Weakneffes, moft Remarkable Effects from it, and have had Succefs beyond expectation; it fupporting the Spirits, and relieving them, where a more active Medicine might over-power them, and yet not have done that Service.

But this *Stone* in *Goats* and *Monkeys* is a Difeafe, and not Natural; as well as the Stone in the Bladder or Kidneys of a Man. *Bontius* (56) therefore obferving the good Effects of the *Bezoar-ftones* bred in thefe *Animals,* argues with himfelf, why thofe in Men, which he finds laminated in the fame manner, might not be of as great an Efficacy; and upon Tryal, he affures us, that they are fo. *Hoc certè compertum habeo, Lapidem in veficâ hominis repertum, urinam & fudores probè ciere, quod tempore ingentis illius peftis, quæ Anno* 1624 *&* 1625. *Leydam, Patriam meam & reliquas Hollandiæ Civitates, miferandum in modum vaftabat, in penuriâ Lapidis Bafaartici, nos exhibuiffe memini, & Sudorificum (aufim dicere) melius & excellentius inveniffe, cum admixtâ Theriacâ, aut Mithridatio, cum Oleo Succini aut Juniperi guttis aliquot.*

We come now to the Third Stage of the *Ductus Alimentalis,* the *Inteftines*; which ferve for the feparating the *Chyle* from the *Fæces,* and fo tranfmitting it into the *Vafa Chylifera,* or *Venæ Lacteæ,* as they are call'd, which conveys it into the *Blood,* for the recruiting the conftant wafte that is made there, and repairing it's loffes; as alfo for the Nourifhment and Augmentation of the Parts: And for the doing this, 'tis requifite that the *Inteftines* fhould be long; and they being fo, that they fhould be coyled and winding; that this *Separation* might be the better performed, and fo we find the *Guts* in our *Pygmie.* For from the *Pylorus*

(53) Jac. Bontij, *Hift. Nat. & Med.* lib. 4. *in Animadv. in Garcia ab Orto,* cap. 45. p. m. 48. (54) *Mifcell. Curiofa German. Decuria fecunda annus fecundus An.* 1583. *Obferv.* 189. p. 420. (55) Jo. Bapt. Tavern. *lib.* 2. *Itin. Indic. cap.* 24. (56) Bontius *ibid.* in cap. 46. p. m. 48.

to the *Anus*, they meaſured Thirteen Feet and three Inches, *viz.* from the *Pylorus* to the *Cæcum* or beginning of the *Colon*, was Nine Foot Ten Inches ; and the *Colon* and *Rectum* were Three Feet and Five Inches long. The *Cæcum* here, or *Appendicula vermiformis*, was Four Inches and three quarters long. So that the length of the *Guts* here, in proportion to the length of the Body, is much the ſame as 'tis in a *Man*. But in two of the *Sapajous* diſſected by the *Pariſians*, the whole Inteſtines were but Five Foot two Inches ; and in the other two *Monkeys*, Eight Foot long. So that herein our *Pygmie* more reſembles a *Man*, than their *Monkeys* did.

And as in the length, ſo likewiſe in other Circumſtances, the *Inteſtines* of our *Pygmie* were liker to thoſe of a *Man*, than thoſe of the *Monkey* and *Ape-kind* are. For the *Pariſians* tell us, that in their *Monkeys*, *the Inteſtines were almoſt all of the ſame bigneſs, and that the* Ileon *was in proportion a great deal bigger, than in a Man.* In our Subject we found a ſenſible difference. For the *ſmall Guts*, which were much of a bigneſs, being a little extended, meaſured in Compaſs about Two Inches and three quarters. The *Colon* was Three Inches and three quarters about ; and the *Appendicula Vermiformis* (which was in our *Pygmie* as 'tis in a *Man*, and is not to be met with in *Apes* and *Monkeys*) was about the bigneſs of a Gooſe-quill. It's length I have mentioned before.

Into the *Duodenum* of our *Pygmie*, a little below the *Pylorus*, were inſerted the *Ductus Communis* of the *Gall*, and the *Ductus Pancreaticus* ; they both emptying themſelves into the Gut at the ſame *Orifice* as is uſual in *Man*. And the ſame is obſerved likewiſe by *Drelincourt* in the *Male Ape* he diſſected, where he tells us, *à Pyloro qui videtur ſuggrunda eſſe circularis & carnoſa principio Ecphyſeos præpoſita, ad foramen uſque intra eandem Ecphyſin Commune Ductui Bilario & Wirzungiano, præciſe pollex eſt Mathematicus ; ab illo autem foramine intra duplicem Ecphyſeos tunicam ſtilus gracillimus intruſus eſt in prædictam Veſiculæ felleæ recurvitatem, rurſusque ab eodem Inteſtinali foramine idem ſtilus compulſus eſt in Ductum Wirzungianum.* But the *Pariſians* obſerved in the *Monkeys*, that *the Inſertion of the* Ductus Pancreaticus *into the Inteſtine,* (*which in Man is always near the* Porus Bilarius *) was Two Inches diſtant there-from.* So that in this Particular the *Monkey* does not ſo much reſemble a *Man*, as *Apes* and our *Pygmie* do.

The Convolutions and Windings of the *ſmall Guts* in our *Pygmie*, and their Situation, were much the ſame, as in a *Man :* And they were all plentifully irrigated with *Blood-Veſſels*. In the inward Coat of the *Inteſtines* I could obſerve the *Miliary Glands*, deſcribed by Dr. *Willis* ; as alſo thoſe larger cluſters of *Glands*, mentioned by *Joh. Conrad. Peyerus.* The *Colon* I thought proportionably longer, than 'tis in a *Man*. It had the ſame *Ligaments* and *Cells*, and leaves of *Fat* hanging to it, as a Man's
hath ;

hath ; and the fituation, was the fame : but it being fo long, it had more windings than ufually. The *Parifians* obferved in their *Monkeys*, that the *Colon was not redoubled like an S. as in* Man, *being quite ftrait.* *Drelincourt's Ape* was more like ours, for fpeaking of the *Colon*, he faith, *retorquetur variè antequàm producat Rectum* ; *cellulas habet ut in Homine.* For the length of the *Colon* in the *Monkeys* diffected by the *Parifians*, was but thirteen Inches ; and an Inch in Diameter ; whereas, the *Colon* of our *Pygmie* with the *Rectum*, was three Foot five Inches, as I have mentioned ; and therefore liker to a Man's , and requiring thefe convolutions the more.

In a *Man* the *Inteftines* are commonly diftinguifhed into *Inteftina Tenuia* and *Craffa :* The *Tenuia* are fubdivided into the *Duodenum*, *Jejunum*, and *Ileon* ; The *Craffa*, into the *Cæcum, Colon* and *Rectum :* and the *Cæcum* commonly is reputed that *Appendicula Vermiformis* , which is placed at the beginning of the *Colon*, where the *Ileon* empties it felf into it. Now this Part in a *Man*, being fo fmall ; and being obferved never to contain any Excrement ; I can't think, that it deferves the Name of an *Inteftine*, much lefs to be reputed one of the *Craffa*. 'Tis true , in *Brutes*, this part is often found to be very large and capacious ; and to be filled with *fæces* ; and in fuch, it may be juftly efteemed an *Inteftine*. As in a *Rabit*, 'tis very long and hath a *Cochlear Valve* ; fo in an *Oftridge*, there are two *Cæcums* ; each a yard long, with a like *Valve*. But in *Man*, 'tis far different. Many therefore do not think this *Proceffus Vermiformis* , to be the *Cæcum* ; but rather take for is, that bunching out of the beginning of the *Colon* ; which is projected beyond the entrance of the *Ileon* ; which in the Common *Ape* and *Monkey* is more, than in a *Man*. However, I think it not enough, as to make it a diftinct *Inteftine*; and the number of the *Inteftines* in a *Man*, ought to be made fewer.

Our *Pygmie* therefore having this *Proceffus Vermiformis* in all Circumftances, fo like to that in a *Man* ; and *Monkeys* and *Apes* having nothing like it : it is a remarkable difference of our Subject from them, and an agreement to the Structure of a *Humane* Body. So the *Parifians* tell us in their *Monkeys, the Cæcum had no Vermiform Appendix.* So in the *Ape* diffected by *Blafius*, he faith , *Proceffus Vermiformis in totum hic defideratur.* And fo *Drelincourt, Cæcum caret Epiphyfi Vermiformi, qualem homines habent.* We will fee therefore, what kind of *Cæcum* 'tis, that they defcribe in the *Monkeys* and *Apes*.

The *Parifians* tell us , in the Diffection of their *Monkeys* ; That *the* Cæcum *was very large, containing two Inches and half in length ; and an Inch Diameter at the beginning : it went pointing, and was fortified by three Ligaments, like as the* Colon *is in* Man; *there to form little Cells* ; *this Conformation is wholly different from that of a Man's* Cæcum. 'Tis true, 'tis more projected, than in a Man ; So *Blafius* in his *Ape* , makes it jutting

F

out

out beyond the Inſertion of the *Ileon*, *Manus tranſverſæ*, *ſeu trium digi-torum ſpatium.* And *Drelincourt* tells us, *duarum unciarum eſt.* But ſince it hath thoſe *Ligaments* of the *Colon*, 'tis plain, that 'tis only a part of it, and not a diſtinct Inteſtine; or as *Blaſius* more truly calls it, *Principium Coli.* He hath given a *figure* of it, but not very exact; and in another *figure* he repreſents the *Valve* of the *Ileon* at the *Colon*, or rather *Valves*; for he makes more than one. His *Deſcription*, as 'tis faultily printed; ſo I am afraid, it is not very accurately drawn up, and therefore do o-mit it.

But what is different from *a Man*, as alſo from the *Ape* and *Monkey* too, or any other *Animal* I yet know of; is a ſort of *Valve* I obſerved at the other Extream of the *Colon* in our *Pygmie*, where it paſſes into the *Rectum.* For the turn of the *Colon* here, is very ſhort; and in the in-ſide I obſerved a Membranous Extenſion like a *Valve*, an Inch in length, which divided the Cavity half way. The *Rectum* did not much differ from the *Colon* in the magnitude of it's *fiſtula*, but was much the ſame; and in other reſpects, as 'tis in a *Man.*

This great length of the *Inteſtines* in our *Pygmie* was orderly colliga-ted and faſtened to the *Meſenterie*, which kept them in a due ſituation; and ſo, as to make in them, ſeveral windings or convolutions; that hereby they might the better make a diſtribution of the *Chyle*; and the whole was, as 'tis in a Man. But I obſerved here, the *Membranes* of the *Meſenterie*, to be more looſly joyned together, than uſually. For by moving them by my fingers, I found the *blood veſſels* which were faſten-ed to the *upper Membrane*, would eaſily ſhove over thoſe, that were fixt in the *under Membrane* of the *Meſenterie*; and run on either ſide of one another, as I would draw them. I have ſometimes ſeen the ſame in *Hu-mane Bodies.*

The *Meſaraic Veſſels* here, were very numerous; as they approach the *Inteſtines*, they form ſeveral *Arches*, whereby they communicate with one another; and from theſe *Arches*, they ſend out numerous Branches to the *Inteſtines* of each ſide, which run claſping them; after-wards they ſubdivide, and inoſculate with one another in infinite *Ramu-li*: ſo that by injecting theſe *Veſſels* with *Mercury*, they appeared ſo nu-merous; as almoſt wholly to cover the Trunk of the *Inteſtines.* And the ſame is in *Man.*

I have likewiſe ſeen, by injecting the *Meſenterick Veſſels*; that the *Mer-cury* has paſſed into the *Lymphæducts*; and ſo into the *Venæ* or *Vaſa La-ctea.* Which is a great contrivance of *Nature.* For the *Motion* of all *fluids* being *Pulſion*, without this advantage, part of the *Chyle*, muſt ne-ceſſarily ſtagnate in ſome of the Veſſels; till a freſh diſtribution of *Chyle* comes; to protrude it on; and ſo it would be apt to coagulate and cauſe

cauſe Obſtructions. But by the *Lympha* thus paſſing into them ; the *Chyle* is ſtill forced forwards, and the *Veſſels* waſhed clean of it ; and being thus often moiſtened, they are preſerved from becoming over dry, or cloſed or obſtructed. So Provident therefore is *Nature*, that in the whole *Via lactea*, not only in the *Meſenterie*; but into the *Receptaculum Chyli*, and *Ductus Thoracicus* likewiſe; abundance of *Lymphæducts* are emptied. Which gives us one good Reaſon ; that *Nature* does not act in vain, in making ſuch a ſeparation of a Liquor from the Maſs of Blood ; which is ſo ſoon to be return'd to it again ; ſince hereby ſhe performs ſo great an Office.

In the *Meſenterie* of our *Pygmie* I obſerved ſeveral ſmall *Glands* ſcattered up and down, as in a *Man*; but not ſo regularly amaſſed together in the middle ; as the *Pancreas Aſellij* is in *Brutes*. And *Drelincourt* obſerved much the ſame in the *Male Ape*. *Glandulæ ad radicem Meſenterij, & paſſim in ambitu, numeroſæ & planæ, magnitudinem lentulæ, ſed Ovales. Anaſtomoſes frequentiſſimæ Venarum cum Venis & Arteriarum cum Arteriis in univerſo Meſenterij circulo.* And as that part of the *Meſenterie* which faſtens the *Colon* is call'd *Meſocolon*; ſo for the ſame reaſon, that ſlip of it repreſented in our figure, that runs down to the *Proceſſus vermiformis*, may be call'd the *Meſo-cæcum.*

We ſhall next proceed to the *Liver*, in which part our *Pygmie* very remarkably imitated a *Man*, more than our common *Monkeys* or *Apes* do. For the *Liver* here was not divided into *Lobes* as it is in *Brutes*; but intire as it is in a *Man*. It had the ſame ſhape ; it's ſituation in the body was the ſame; and it's Colour, and Ligaments, the ſame. It meaſured in it's greateſt length about five Inches and an half; where broadeſt, 'twas about three Inches; and about an Inch and three quarters in thickneſs. Towards the *Diaphragm* 'twas *convex* : it's under part was *Concave*, where it receives and emits the Veſſels, having a little *Lobe* here, as 'tis in a *Man*.

The *Pariſians* remark in the *Monkeys* they diſſected, that *the Liver was very different from the Liver of a Man, having five Lobes as in a Dog*; viz. *two on the right ſide*; *and two on the left*; *and a fifth laid upon the right part of the body of the Vertebræ. This laſt was divided, making as it were two leaves.* So *Drelincourt* in the *Male Ape* obſerves, *Jecoris Lobi duo juxta umbilicalem venam, quorum ſecundo incunenta erat veſicula fellis, duo alij ventriculum amplectebantur, cum lobulo quinto ſe inſerente in ſpatium ventriculi intra orificium utrumque.* So likewiſe in the *Female Ape* he tells us, *Jecur opplet regionem Epigaſtricam quintuplici lobo, uno ſexto minimo opplens cavitatem lunarem ventriculi.* But *Blaſius* in the *Ape* he diſſected ſaith, *Epar cum humano minimè, optimè cum Canino convenit, maniſeſtiſſimè in lobos VII diviſum, tantæ magnitudinis ut etiam utrumque Hypochondrium*

F 2   *drium*

drium occupet.. Veſalius (57) therefore is in the right; where he ſaith, *Quæ enim Diſſectionum Profeſſores de Jecoris formâ, ac penulis ſeu fibris (quos λόβες; Græci vocant) commentantur ; è Canum potiùs , & ſimiarum ſectio-nibus, quàm hominum didicerunt. Humanum enim Jecur in fibras, Porcini, ac multò adhuc minùs Canini Jecoris modo, non diſcinditur.* And that he hints here at *Galen,* is plain, from what he expreſſes in his Epiſtle *ad Joachim Roelants,* (58) where he farther enlarges upon it. And *Galen* (59) himſelf tells us, that *Herophilus* was of this Opinion. So *Theophilus Protoſpatarius* (60) ſaith, that the *Liver* is divided into four *Lobes* ; and gives us there a diſtinct Name for each. *Ariſtotle,* (61) 'tis certain, was much more in the right, where he ſaith, ςρογζύλον δ᾽ ἐςὶ τὸ τῷ ἀνθρώπȣ ἦπαρ, ἢ ὅμοιον τῷ βοεῖῳ. *Rotundum Jecur hominis eſt, ac ſimile bubulo.* For the Liver of a Bullock, like a Man's is entire ; and not divided into Lobes. However *Franciſcus Puteus* (62) in his *Apology*, having named ſeveral Phyſicians and Chirurgians, that were with him at the opening of *Charles* the Ninth, Duke of *Savoy*, ſaith, *hi omnes ut Jovem mihi poſ-ſunt eſſe Teſtes, quod obſervatum eſt Epar habuiſſe quatuor pinnulas. Jaco-bus Sylvius* (63) likewiſe juſtifies *Galen*, againſt *Veſalius* ; and tells us , *Quin & Hippocrates Lobos Epatis humani quinque connumerat libro ſuo de oſſibus. Rufus autem quatuor vel quinque.* But *Renatus Henerus* (64) hath anſwered *Sylvius* as to this matter ; and there needs no farther diſpute a-bout it, if one will but believe his own Eyes, he may fully ſatisfie him-ſelf, that, in an *Humane Liver* there are none of thoſe *Lobes*, but that 'tis one entire Body ; as it was alſo in our *Pygmie*. But . in *Apes* and *Monkeys* the Liver is divided into *Lobes.*

The great *uſe* of the *Liver* is for to make a ſeparation of the *Gall* from the Maſs of Blood. We will therefore here examine the *Biliary Veſſels* ; nor do I find them any thing different from thoſe in a *Man* ; only the *Bladder* of *Gall* here in our *Pygmie* ſeemed longer, being four Inches in length. It's adhæſion to the *Liver* was not ſo much as it is in a *Man* ; for at the *fundus* or end, it juts beyond the *Liver* about half an Inch. For about three quarters of an Inch, it is more cloſely joyned to the *Liver* ; afterwards it is faſtened to it only by a Membrane, as is alſo the *Ductus Cyſticus*. So that the *Veſica fellea* when inflated with wind, ſeemed more to repreſent an *Inteſtine* by it's *anfractus* and length , than the uſual ſhape of the *Bladder of Gall* ; which commonly is more belly-ing out.

The *Pariſians* obſerved in their *Monkeys*, that the *Bladder was faſtened to the firſt of the two Lobes which were on the right ſide. That it was an Inch long, and*

(57) *Veſalij de fabricâ corporis humani*, lib. 5. cap. 7. p.m.619.   (58) *And. Veſalij Epiſtola*,&c.p.m.81.
(59) *Galen. de Anat. Adminiſtrat.* lib.6.cap.8.. (60) *Theophilus de Corporis humani fabrica*, lib. 2. cap. 2.
(61) *Ariſt. Hiſt. Animal.* lib. 1. cap. 17.p. m. 555.   (62) *Franc. Putei Apologia pro Galeno in Anatomicis contra Andr. Veſalium*, lib. 5. p. m. 153.   (63) *Vaſani cujuſdam Calumniarum in Hipp. Galenique rem A-nat. depulſio. per Jac. Sylvium.* vid. Depulſ. 26. p. m. 150.   (64) *Renat. Henerus adverſus Jacobi Sylvij Depulſionum Anat. Calumnias pro Andrea Veſalio Apologia*, p.m. 55.

half

half an inch broad; it had a great Ductus, which was immediately inserted underneath the Pylorus. This Ductus received three others, which instead of that, which in Man is single, and which is called Hepaticus; these three Ductus's had their Branches dispersed like Roots into all the Lobes of the Liver, so that the first had four roots, viz. one in each of the three right Lobes, and one in the first of the left; the second and third Ductus had both their roots in the second of the left Lobes, these branches did not run under the Tunicle of the Liver, so that they were apparent, and not hid in the Parenchyma, as they generally are. But in our Subject the distribution of the Ductus Hepaticus was altogether the same as it is in Man. In the Male Ape, Drelincourt describing the Bladder of Gall, saith, Vesicula fellea longa 2½ pollicibus à fundo ad cervicem, ubi recurvitatem habet maximam, dimidiatè hæret mersa substantiæ Jecoris.

The Ductus Hepaticus in our Pygmie issued out of the Liver with two branches; one arising from the right, the other from the left part of the Liver; and after a short space, joined into one Trunk; and that, after a little way, joyning with the Ductus Cysticus, do form the Ductus Communis, which empties it self into the Duodenum a little below the Pylorus, at the same Orifice with the Ductus Pancreaticus, exactly as 'tis in Man, as I have mentioned.

At the Simous part of the Liver I observed the Vena Porta to enter, as likewise the Epatic Arteries and Nerves. And here in the Membrane about these Vessels, I observed a pretty large whitish Gland. The Vena Umbilicalis entered the Liver at the fissure. It seemed large, but I found it's fistula or pipe was closed. The Vena Cava issued out of the Liver at the Convex part, where 'twas joyned to the Diaphragm.

In the Spleen of our Pygmie I did not observe any thing extraordinary, or different from a Humane Spleen. It was of a lead Colour, and of the shape represented in our figure; 'twas fastened by Membranes to the Peritonæum; and by the Omentum and Vasa brevia to the Stomach, so that upon inflating the Stomach, the Spleen would be brought to lye close on the Stomach, as if it was fastened immediately there. The Spleen here was two Inches and an half long; and one Inch and a quarter broad; and seated as usually in the left Hypochondre under the Bastard Ribs. The Ramus Splenicus was very remarkable, sending it's Trunk along the Pancreas, as in Man, and having numerous branches near the Spleen.

The Parisians tell us, that in their Monkeys the Spleen was seated along the Ventricle as in Man; but it's figure was different, in one of our Subjects being made as the Heart is represented in Blazonry; it's Basis containing an Inch. They give a figure of it, but nothing like that of ours, which more represented the figure of an Humane Spleen; tho' in Man it's figure is often observed very different. Blasius in the Ape he dissected, observes
that

that the Spleen *triangularis figuræ est, exiguus admodùm respectu corporis, coloris nigricantis, læve equidem molleque valdè corpus, ast exteriùs inæquale, quasi ex globulis variis confectum, adeò ut etiam conglomeratis Glandulis Substantiam Lienis annumerare velle, tali in subjecto fundamentum aliquod agnoscat. Ex Ramo Splenico numerosos eosque insignes Ventriculo suppeditat ramos, magnitudinem & figuram externam Fig. 3ᵃ. Tab. XI. exhibet.* But his *figure* of the *Spleen* was nothing like to that of ours. For I did not observe those inequalities in the *superfice* which he represents in his, to exhibit the *conglomerate Glands.* 'Tis true , having injected the Spleen of our *Pygmie*, the *Mercury* filling the *cellulated* body of the *Spleen*, did make an appearance on the surface somewhat like those inequalities in his *figure.* But *Frederic. de Rusch* (65) is very positive, that neither those *Glands*, nor *Cells* mentioned by *Malpighius*, are to be met with in a *Humane Spleen :* tho' he grants, that they are in the *Spleen of Brutes. Drelincourt* in the *Female Ape* saith, *Lien Scalenum figura refert, cohæret Reni sinistro & liber est à Diaphragmate.* And in the *Male Ape* he observes, *Lien triangularis & crassior quàm in fæminâ, Pancreas excipiens.*

We shall therefore now proceed to the *Pancreas*, which in our *Pygmie* was situated, just as it is in a *Humane Body* ; lying under the *Stomach*,transverse to the Spine,from the *Spleen* towards the *Liver.* It was about two Inches long, about half an Inch broad, of a white yellowish Colour ; it's surface uneven, being made up of abundance of *Glands* ; it's *Secretory Duct* emptied it self into the *Duodenum*, just where the *Ductus Communis* of the *Gall* doth, as I have mentioned before.

The *Parisians* in their *Monkeys* observed, that the Pancreas *had only it's figure, which made it to resemble that of Man* ; *it's connection*, and *insertion being wholly particular.* For it was *strongly fastened to the Spleen* ; and the *insertion of it's* Ductus *into the Intestine* (*which in Man is always near the* Porus Bilarius) *was two Inches distant therefrom. Blasius* in his *Ape* describes it thus ; *Pancreas ventriculo substratum, solidæ admodùm substantiæ est, nec adeò molle, quàm in Canibus aliisque Animalibus notatur. Longum itidèm insigniter, ast latitudinis ejus, quæ nè minimi digiti latitudini respondeat.*He takes no notice here,how the *Ductus Pancreaticus* was inserted ; which *Drelincourt* tells us in the *Female Ape* was eight lines above the *Porus Bilarius. Pancreas connatum Lienali Caudæ, & extremo Reni sinistro. Ejus ductus inscritur octo lineis supra Porum Bilarium , contrà ac Canibus, substernitur immediatè Ventriculo, & supersternitur brevi Intestino.* Tho' in the *Male Ape* he tells us, 'tis inserted into the *Duodenum* at the same Orifice with the *Duct* of the *Gall* ; as I have already mentioned and quoted before.

---

(65) *Epistola Anatomica Problematica quinta.*

i   **The**

The *Glandulæ Renales* in our *Pygmie* were very large, and placed a little above the *Kidnies* as they are in *Man*. That on the right side, was of a triangular; that on the left of an oblong figure. They were about three quarters of an Inch long: and almost half an Inch broad. They had the same Vessels, as there are in a Man.

The *Parisians* in their *Monkeys*, observe that the *Gland called* Capsula Atrabilaria, *was very visible, by reason that the Kidney was without fat. This Gland was white, and the Kidney of a bright red*; *it's figure was triangular*. *Blasius* in his *Ape* tells us, *Glandula Renalis triangularis ferè figuræ est, notabilis valdè pro ratione Corporis*, and gives us a *figure* of it, which was nothing like ours. What *Drelincourt* remarks in the *Male Ape*, is, *Capsulæ Atrabilariæ triplicem Scrobiculum habent, quarum liquor expressus linguam non ità constringit, uti in Capsulis fœmineis*. And in the *Female Ape* he tells us, *Ren Succenturiatus sinister ab Emulgente venam habet*; *idem major Dextro*. This I observed in our *Pygmie*; but he saith nothing farther here of their *Liquor*, nor did I taste it in ours.

We shall now proceed to the *Kidneys*. In our *Pygmie* I did observe very little or no fat in the *common* or outward *Membrane*, usually called *Adiposa*: *Drelincourt* observed the same, *nullus hic Adeps in Tunicâ communi vel propriâ*, as he tells us of his *Ape*. The *Kidneys* of our *Pygmie* were two Inches and a quarter long, an Inch and an half broad; and about an Inch in depth. They had not altogether so large a *Sinus* at the Entrance of the *Emulgent* Vessels, as there is in a Man's; and the whole appeared somewhat rounder; but their situation was the same, as were likewise the *Emulgents*. Having divided the right *Kidney* length-ways, I observed the *Cortical* or *Glandulous* Part to appear like a distinct Substance, being a of tawny or yellowish colour; and different from the *Inward* or *Tubulary Part*; which was more entire and compacted together, than in a *Man's*; and was of a red colour, by means of the *blood vessels* which run between the *Tubuli Urinarij* or *Secretory Ducts*, which make up this part of the *Kidneys*. Which Vessels when inflamed and overextended, by making a Compression on these *Tubuli Urinarij*, may cause a *Suppression* of Urine; in which case *Phlebotomy* or *Bleeding* is very necessary. And without doubt was the Cause of the Success *Riverius* (66) met with in a Patient, who had a *Suppression* of Urine for three days; for upon bleeding freely, he was presently relieved, and in a short time rendered a large quantity of Urine. In this *Tubulary* Part of a Humane *Kidney* I always observe these *Blood Vessels*: but here usually the *Cortical* or *Glandulous* Part makes a deeper descent between the heads of this *Tubulary*, and divides it into several Bodies; and as many of them as appear, so many lesser *Kidneys* may be reckoned to make up the Body of each *Kidney*. In *Infants* the *Kidney* externally appears more divided

(66) *Riverij Observ. Med. Cent.* 1. *Obs.* 1.

than

than in *Adult* Perſons; but moſt remarkably they are ſo, in a *Bear*, the *Porpois* and an *Oſtrich*; where there are abundance of diſtinct *ſmall Kidneys* amaſſed together to make up each.

The *Pariſians* in the *Kidneys* of their *Monkeys* obſerve, *that they were round and flat; their ſituation was more unequal, than in a Man; the right being much lower than the left*, viz. *half it's bigneſs.* *Drelincourt* in the Female Ape remarks, *Renes globoſi, dexter intra Hypochondrium incumbit Coſtæ* 11. *&* 12. *Siniſter locum habet intra Coſtam ultimam. Altitudo Renis dimidiæ unciæ. Renalium venarum dextra longè elatior ſiniſtrâ. Rene aperto Carnis diſcrimen ut in homine, exterior quidem nigricans lineis quatuor craſſa, interior albicans lineis duabus.*

The *Pelvis* of the *Kidney* in our *Pygmie* was as 'tis uſually in a *Man*; and the *Ureters* had nothing remarkably different in their Structure, from the common make. They were about the bigneſs of a Wheat ſtraw; and were inſerted into the neck of the *Bladder*, as repreſented in our figure; rather ſomewhat nearer the neck, than in an *Humane Bladder*.

The *Pariſians* and *Blaſius* have no remarks upon the *Ureters*. *Drelincourt* in the *Male Ape* ſaith, *Ureteres ſuprà Pſoas Muſculo & Iliaco, atque ſubtùs vaſis Spermaticis, quibus decuſſatim ſubſtrati ſunt etiam quibuſve vaſcula admittunt, ſeſe reflectunt in Hypogaſtricam, decuſſantes ramos Iliacos & Ejaculatorios.* And in the *Female*, *Ureteris expanſiones arcuatim reflexæ ut in homine. Vaſa habent ſupernè à Renalibus, infernè à Muſculis.*

The *Bladder* of *Urine* in our *Pygmie* was of an Oblong figure, not ſo globous as commonly in *Man*, for being moderately blown up it meaſured four Inches in length; and two Inches and half in breadth. In other Circumſtances 'twas agreeable enough with an *Humane Bladder*.

The *Pariſians* tell us, that in the *Female Monkey*, *the Neck of the Bladder had it's hole otherwiſe than in Women, being very far in the Neck of the Matrix,* viz. *towards the middle, at the place where it's roughneſs began, which were ſeen only towards the Extremity of the Ductus, near the internal Orifice.* *Blaſius* ſaith nothing of it in his *Ape*; and all that *Drelincourt* tells us is, *Veſica Peritonæo ſuſpenſa ut in aliis Brutis.*

Before we proceed to the Parts of Generation (which remain beſides to be here deſcribed) we ſhall a little take notice of thoſe large *Canales* of the *Blood*, the *Arteria Aorta* and the *Vena Cava*, and the *Rivulets* they emit or do receive; all which I find in our *Pygmie* to be juſt the ſame, as they are in a *Man*. For from the *Aorta* ariſes here, the *Arteria Cæliaca*; the *Arteria Meſenterica ſuperior*; then the *Emulgent Arteries*; below them, the *Spermatick Arteries*; then the *Arteria Meſenterica inferior*; then the
Trunk

Trunk divides into the *Iliac Branches*. So the *Vena Cava* too in our *Pygmie* exactly imitated that in a *Man.*

How the Structure of these *Vessels* are in *Monkeys*, the *Parisians* do not tell us, and their *figure* is very imperfect; as is likewise that in *Blasius*, which seems altogether fictitious. What he writes, is this; *Arteria magna circa Renem dextrum succumbit Venæ Cavæ, & ubi Iliacos Ramos constituit eandem supergreditur; contrà ac in Homine, Cane, aliisque animalibus fieri reperimus, ubi sinistra occupat, hinc à sinistra ad dextram progreditur supra Arteriam.* So *Drelincourt* tells us in the *Male Ape*, *Aorta descendens mox atque bifurcatur equitat, & adscendenti Cavæ incumbit.*

We come now to the *Parts of Generation*; and shall begin with the *Vasa Præparantia*; The *Arteries* and *Veins*. The *Spermatic Arteries* in our *Pygmie* do both arise out of the Trunk of the *Aorta*, a little below the *Emulgent Arteries*, as in our *figure*; and after having ran a little way, they meet with the *Spermatic Vein*; and are both included in a common *Capsula*, and so do descend to the *Testes*. These *Arteries* do carry the *blood* to the *Testes*, from whence the *Semen* is afterwards separated; the residue of the *Blood* is return'd from the *Testes* by the *Spermatic Veins*; whereof that on the right side enters into the Trunk of the *Vena Cava*, a little below the right *Emulgent Vein*; and that of the left, is emptied into the left *Emulgent Vein*, just all one as it is in a *Humane Body*. Having injected the *Spermatic Vein* with *Mercury*, it discovered abundance of Vessels, running waving; which otherwise did not appear: and a great many of them were extreamly fine and small.

The *Parisians* give no description of the *Spermatic Vessels* in their *Monkeys*; and in their *figure* the left *Spermatic Vein* is omitted, or left out. *Thomas Bartholine* (67) in his *Anatomy* of a *Mammonet* (which he describes, as not having a Tail; and therefore it must be of the *Ape*-kind, and not a *Cercopithecus*, or a *Monkey*, as he calls it) in his figure of these parts, represents the left *Spermatic Vein*, emptying it self into the left *Emulgent*, as it is in our Creature. *Blasius* therefore in the account of the *Ape* he dissected, must be mistaken; both in his figure and description too; for in the former, he represents the *left Spermatic Vein* running into the Trunk of the *Cava*; and justifies it in the latter; in telling us, *Vasa Spermatica utroque latere ex Trunco Cavæ & Aortæ oriuntur, & quidem altiori loco ea quæ sunt lateris dextri, inferiore quæ sinistri.* But *Drelincourt* certainly is more in the right, who informs us, that in the *Male Ape* he dissected, *Vena Spermatica dextra crassa, & ab interiori trunco Cavæ adscendentis pollice infra Emulgentem sinistram enascitur, surculosque emittit sinistros in Membranas vicinas. Arteria Spermatica dextra à trunco anteriori*

(67) *Thom. Bartholin. Acta Medica & Philos. Hafnions. an.* 1691. & 1672. Obs. 36.

G

Aortæ paulò infra Emulgentem ſiniſtram enaſcens ſub Venâ Emulgente inter-
cruciat Cavam aſcendentem, quæ ſuperinequitat, & conjungitur Venæ Conſo-
ciali cò præcisè loci ubi Vena inſeritur ſuum in truncum. Siniſtra Vena Sper-
matica inſeriturin Emulgentem juxta truncum Cavæ , & conſocialem Arteri-
am admittit eò præcisè loci, in quo enaſcitur dextra. So in the *Female Ape*
he faith, *Spermatica Vena ſiniſtra ab Emulgente ſiniſtrâ, dextra è Trunci*
parte anteriore, pollice infra Emulgentem ſiniſtram.

We come now to deſcribe the *Teſtes*, which in our *Pygmie* were not
contained in a *pendulous Scrotum*, as they are in *Man*, but more contracted and purſed up by the outward Skin, nearer to the *Os Pubis*, and were
ſeated by the ſides of the *Penis*, without the *Os Pubis* ; and I obſerved
them bunching out there , before the Diſſection ; ſo that it ſeemed to
want a *Scrotum* ; or at leaſt the Skin which incloſed them, was not ſo dilated, as to hang down like a *Cod* ; but contracted them up nearer to the
Body of the *Penis* ; which to me ſeems a wiſe Contrivance of *Nature.*
For hereby theſe Parts are leſs expoſed to the injuries, they might otherwiſe receive in climbing Trees, or other accidents in the Woods. However, the outward Skin here that incloſes them , performs altogether the
office of a *Scrotum.*   And if I miſtake not, I obſerved that *Sepimentum*,
as in a *Humane Scrotum* ; which is made by a deſcent of a *Membrane* there,
which divides each *Teſticle* from one another.

But whether the *Teſtes* being thus cloſely purſed up to the Body, might
contribute to that great *ſalaciouſneſs* this *Species* of *Animals* are noted for,
I will not determine : Tho' 'tis ſaid, that theſe *Animals* , that have their
Teſticles contained within the Body, are more inclined to it, than others.
That the whole *Ape*-kind is extreamly given to *Venery*, appears by infinite Stories related of them.   And not only ſo, but different from other
Brutes, they covet not only their *own Species* , but to an Exceſs are inclined and ſollicitous to thoſe of a *different*, and are moſt *amorous* of
fair *Women.* Beſides what I have already mentioned , *Gabriel Clauderus*
(68) tells us of an *Ape,* which grew ſo amorous of one of the *Maids* of
Honour, who was a celebrated Beauty, that no Chains, nor Confinement,
nor Beating, could keep him within Bounds ; ſo that the *Lady* was forced to petition to have him baniſhed the Court.   But that Story of *Caſtanenda* in his *Annals* of *Portugal* (if true) is very remarkable ; of a Woman who had two Children by an *Ape.* I ſhall give it in *Latin,* as 'tis
related by *Licetus* ; and 'tis quoted too by *Anton. Deuſingius* (69) and
others.

In hanc Sententiam faciunt (ſaith Fort. Licetus (70) *) quæ Caſtanenda*
retulit in Annalibus Luſitaniæ *de filiis ex muliere, ac ſimio natis, mulierem*

---

(68) *Miſcell. Curioſa German. Decur. 2. Ann. 5. Obſ.* 187.   (69) *Ant. Deuſingij Faſcicul.Diſſertat.ſelect.de*
Ratione & Loquela Brutorum, p. m. 196.   (70) *Fortun. Licetus de Monſtrorum Cauſis,*lib.2.cap.68.p.m.217.
nempe

nempe ob quoddam crimen in insulam defertam navi deportatam , quum ibi exposita fuisset, eam simiorum, quibus fertilis locus erat, agmen circumstetisse fremebundum ; supervenisse unum grandiorem , cui reliqui loco cesserint : hunc mulierem blandè manu captam in antrum ingens abduxisse, eique cum ipsum tum ceteros copiam pomorum, nucum, radicumque variarum apposuisse; & nutu ut vesceretur invitâsse ; tandem à ferâ coactam ad stuprum ; facinus hoc multis diebus continuatum adeò, ut duos ex ferâ liberos pepererit : ita miseram (quantò mors optabilior !) victitâsse per annos aliquot ; donec Deus misertus navim eò Lusitanam detulisset ; quumque milites in terram aquatum ex proximo ad antrum fonte exscendissent : abessetque fortè fortuna simius ; feminam ad invisos diu mortales accurrisse, & occidentem ad pedes supplicâsse, uti se facinore, & calamitosissima servitute irent ereptum, adsentientibusque casum, miserantibus illis, eam cum ipsis navim adscendisse. Sed ecce tibi simium supervenientem inconditis gestibus , & fremitibus conjugem non conjugem revocantem : ut vidit vela ventis data, concito cursu de liberis unum matri ostentat, minatur, ni redeat, in mare præcipitaturum ; nec segniter fecit, quod minatus : tum recurrit ad antrum, & eâdem velocitate ad littus rediens ostentat alterum, minatur, & demergit : subsequitur, donec undæ natantem vicere. Rem totam Lusitania teste notissimam, & à Rege mulierem Ulyssipone addictam ignibus, quorundam precibus vita impetrata, lethum cum claustro perpetuo commutâsse.

But to return to our Business. Our *Pygmie* in this Particular of the *Scrotum*, more resembles the *Ape*-kind, than a *Man*. For the *Parisians* tell us, that *the Parts of Generation in three of our Subjects, which were Males, were different from those of Man, there being no* Scrotum *in two of these Subjects, and the Testicles not appearing, by reason that they were hid in the fold of the Groyne. It is true that the third, which was one of the Sapajous, had a Scrotum, but it was so shrunk, that it did not appear.* Or, as they afterwards express it, *The Testicles were shut up in a* Scrotum, *which joyned them close up to the Penis.* So in the *Ape Blasius* describes, *Testes insignes satis, sacculo suo inclusi, non dependent extra abdomen, ad modum eum quo in Homine, Canibus, similibusque Animalibus aliis, sed vicini adeò sunt tendinibus musculorum Abdominis, quos vasa Spermatica transeunt, ac si iis uniti essent, sic ut potiùs in Inguine utroque collocatos eos dicerem,quàm ultra ossa Pubis a Corpore pendulos.* And so *Drelincourt* to the same purpose ; *Scrotum pendulum nullum est, sest Testiculi utrinque juxta Ossis Pubis summa latera, vel Spinam summam ejus decumbunt extra prorsus Abdominis cavum, & proindè extra Musculorum Epigastrij Aponeuroses.*

In the other Parts I am here to describe, I find our *Pygmie* more conformable to the Structure of the same in a *Man*. For the *Testes* were included in a *Tunica Vaginalis*, and had a *Cremaster Muscle* ; which being separated, I observed the *Epididymis* large, and the Body of the *Testis* to be about the bigness of a *Filbird* ; and it's compounding Parts nothing at all different from those of a *Man*. *Jacobus Sylvius*
G 2   *vius*

vius (71) in the *Ape* he diſſected, obſerved, the *Teſtes humanis majores.*

The *Pariſians* tell us, that in ſome of their Subjects the *Teſticles were long and ſtrait, and but one line in breadth, and eight in length.* In one of their Sapajous *they were found of a figure quite contrary, and almoſt as remote from the figure of thoſe of Man, being perfectly round.* Drelincourt's account in his *Ape* is, *Tunica Elytroides fibris carneis à Cremaſtere conſperſa, ut in homine. Arteria Spermatica miro luſu, ſpiralim revolvitur ſuper Teſticuli dorſum. Teſticulus autem Ventri Epididymidum adhæret, niſi fibrillis paucis & laxis, capite ſuo, quo Spermatica Deferentia admittit, ſeparatur illæſus, cauda autem ſua, qua ejaculatoria vaſa emittit, tot punctula candicantia exhibet, divulſus ab Epididymide, quot à Teſticulo canaliculi protenduntur.*

From the *Epididymis* in our *Pygmie* (as it is in a *Man*) was continued the *Vas Deferens* ; a ſlender *Ductus,* which conveys the *Semen* from the *Teſticle* to the *Veſiculæ Seminales.* Theſe *Veſiculæ* were two *cellulated Bladders* placed under the neck of the *Bladder* of *Urine* ; which on the outſide, did ſeem (as it were) nothing elſe but the *Vas Deferens* dilated, and placed in a waving figure there. And as the Body of the *Teſtes* was made up of a curious convoluted Contexture of *Seminal Veſſels* , which running into fewer, form at laſt the Body of the *Epididymis* ; and *theſe Veſſels* afterwards paſſing all into one *Duct,* do make up the *Vas Deferens :* ſo this *Vas Deferens* here, being dilated and enlarged, does form the *Veſiculæ Seminales.* And the ſame is in a *Man.*

The *Pariſians* here do take notice of that Paſſage in *Ariſtotle* I have already quoted, where he likens the *Parts* of *Generation* in the *Male Ape* to thoſe of a *Dog,* more than a *Man.* But the *Philoſopher* herein, is under a Miſtake ; for , as they inſtance, in the *Penis* of a *Dog* , there is a *Bone,* which is not in the *Monkey's* ; ſo likewiſe in *Monkeys,* there are *Veſiculæ Seminales,* which are not to be met with in a *Dog.* They deſcribe them in their *Monkeys* thus : *The Glandulous Proſtatæ were ſmall* ; *the Paraſtatæ Cyrſoides were in requital very large* ; *they contained an Inch in length* ; *their breadth was unequal , being four lines towards the neck of the Bladder, and a line and an half at the other end, differing herein from thoſe of Man , who has them ſlendereſt near the neck of the Bladder. They were compoſed of ſeveral little Baggs, which opened into one another. The Caruncle of the Urethra was ſmall, but very like to that of a Man.* Blaſius hath given us a *figure* of theſe Parts , which I do not like ; as neither that of the *Pariſians.* He deſcribes them thus: *Veſiculæ Seminales hic valdè amplæ, quæ immiſſo flatu per ductum Seminalem Ejaculatorium inſigniter intumeſcunt. Quod ſi premantur, manifeſtiſſimè obſervamus Mate-*

(71) *Jac. Sylvij Variorum Corporum diſſect. Operum,* p. m. 130.

riam

*riam iis contentam moveri in Meatum Urinarium, Veficæ continuum, &
quidem per foramen fingulare, quod in unoquoque latere unicum eſt, quæ res
occaſionem videtur dediſſe Jacobo Sylvio duos ductus Seminales in ſimiâ con-
ſtituendi.* All that *Drelincourt* faith of them is, *Vaſa ejaculatoria retrò Ve-
ſicam tendunt in Corpuſcula prædura mirè anfractuoſa, ut & ipſum initium
Epididymidis.* Which is very conformable to what I obſerved in our
Pygmie. •

Between the root of the *Penis*, and neck of the *Bladder*, is ſeated the
Corpus Glanduloſum, or the *Proſtatæ*, which in our *Pygmie* appeared the
ſame as in *Man*. The *Pariſians* tell us in their *Monkeys* that they were
ſmall. *Blaſius* in his *figure*, befides the *Proſtates*, which he faith are
Glandula veſicis adſtans, albidior ſolidiorque repreſents another, at the
Letters (H. H.) viz. *Glandula alia, major, rubicunda & plexu Nervorum,
aliorumque vaſorum prædita*; which is no *Gland*, but the *Bulb* of the *Penis*. *Drelincourt* in his *Ape* tells us, *Corpora Glanduloſa duos velut Nates
circa veſicæ cervicem ſuprà Sphincterem exhibent.*

We come now to the *Penis*, which in our *Pygmie* was two Inches
long; the girth of it at the root was an Inch and a quarter; but it
grew taperer towards the end. It had no *frænum*, ſo that the *Præpuce*
could be retracted wholly down; and herein our *Pygmie* is different
from a *Man*. The Slit of the *Penis* here was perpendicular as in a *Man*.
In the *figure* the *Pariſians* give us, it ſeems to be horizontal, as it is
plainly repreſented by *Bartholine* in his *third* and *fourth figure* of his *Mamomet*, altho' by his *ſecond figure* one would think otherwiſe. Whether there was any *Balanus* or *Glans* in the *Penis* of our *Pygmie*, or what
it was, I am uncertain: I do not remember I obſerved any. In my
third figure the *Penis* is repreſented decurtaxed at the end, and without
the *Præputium*, which was left entire to the Skin. *Dreclincourt's* account
of it in the *Ape* is this; *Genitale prorſus expers eſt frænuli ac proinde Præputium devolvitur ad radicem uſque Penis, & denudatur Glans ipſa, atque
Penis integer. Balanus conſimilis virili, excepto frænulo, atque præterea
hiatum maximum exhibet, quâ parte Ligamenta Cavernoſa deſinant, & Glans
utrinque prominet.* At the root of the *Penis* of our *Pygmie*, we obſerved the *Muſculi Erectores* to be ſhort, and thicker proportionably than in
a *Man*; and the *Ligamentum Suſpenſorium* larger: The *Muſculus accelerator Urinæ* was large, covering the *Bulb* of the *Cavernous body* of the
Urethra. The *Corpora Nervoſa*, or the two *Cavernous bodies* of the *Penis*
were divided length-ways by a *Sepimentum* in the middle, as in *Man*.
In the *Urethra* likewiſe there was a *Cavernous body*. The *Veſſels* of the
Penis anſwered exactly to thoſe of a *Man*.

Blaſius in his *Ape* faith, *Penis Nervoſum Corpus unicum tantum habere
videtur, ſepimento notabili deſtitutum.* But I am apt to think he might be
miſtaken; for in our Subject 'twas very plainly divided, but more remarkably

markably towards the root than forwards. What he adds afterwards , *Circa radicem Penis Tuberculum exile occurrit, exteriùs carnoſæ naturæ, interiùs reticulari vaſorum plexu refertum, interſtitia ipſius materiâ rubicundâ occupante,* by this I ſuppoſe he means the *Bulb* of the *Penis*. *Drelincourt* expreſſes it better, where he ſaith , *Totus Penis duobus Ligamentis Cavernoſis à tuberibus Iſchij gaudet.* In our Subject theſe two bodies were very large and *cavernous* within. But what *Drelincourt* adds, *Urethra planè carnoſa* ; This was different in our *Pygmie* ; for as I have mention'd, the ſides of the *Urethra* here were *Cavernous* too , tho' not much.

How the *Organs* of *Generation* are in the *Female* of this *Species* of *Animals,* I have had no opportunity of informing my ſelf. But by *Analogy* I can't but think, they muſt be very like to thoſe of a *Woman,* ſince they are ſo even in *Monkeys* and *Apes* in ſeveral reſpect ; tho' in ſome, they imitate the Structure of theſe Parts in *Brutes*. Thus the *Pariſians* obſerve , *The generative Parts of the Female had alſo a great many things which rendered them different from thoſe of Bitches, herein reſembling thoſe of* Women ; *there were ſome of them likewiſe which were as in* Bitches, *and after another manner than in* Woman ; *for the exteriour Orifice was round and ſtrait, as in* Bitches, *and the generality of other Brutes, and had neither* Nymphæ *nor* Carunculæ. *The Neck of the Bladder had it's hole otherwiſe than in* Woman, *being very far in the Neck of the* Matrix, *viz. towards the middle, at the place where it's roughneſs began, which were ſeen only towards the extremity of the* Ductus *near the Internal Orifice. The Trunks of the* Matrix *were alſo different from thoſe of* Women, *and reſembling thoſe of Brutes in that they were proportionably longer , and more redoubled by various turnings. The* Clitoris *had ſomething more conformable to that which is ſeen in other Brutes that have it, than in that of* Women, *being proportionably greater, and more viſible than it is in* Women. *It was compoſed of two Nervous and Spongious Ligaments, which proceeding from the lower part of the* Os Pubis, *and obliquely advancing to the ſides of theſe Bones, did unite to form a third Body, which was ten lines in length. It was formed by uniting of the two firſt, which a very ſtrong Membrane joyned together, going from one of the Ligaments to the other , beſides a hard and nervous Membrane which inveloped them. They terminated at a Gland like to that of the Penis of the Male. The little Muſcles, which were faſtned to theſe Ligaments, proceeded as uſual from the tuberoſities of the* Iſchium. *Theſe Ligaments were of Subſtance ſo thin and ſpongious , that the wind penetrated, and made them eaſily to ſwell, when blown into the Network of the Veins and Arteries which is in this place. This Network was viſible in this Subject, being compoſed of larger Veſſels than they proportionably are in* Women. *It was ſituated as uſually under the ſecond pair of Muſcles of the* Clitoris. *It's figure was Pyramidal, ending from a very large Baſis in a point, which run along the third Ligament to it's extremity towards the Gland.*

The

The reſt of the *Parts of Generation* were like to thoſe of *Women*. The *Neck* of the *Bladder* had it's *Muſcles as in Women :* For there were a great number of fleſhy *Fibres*, which proceeding from the Sphincter *of the* Anus, were faſtened to the ſides of the *Neck* of the Uterus *, and other ſuch like Fibres* which did come from the Sphincter *of the Bladder to inſert themſelves at the ſame place. The Body of the* Uterus, *it's Membranes, internal Orifice, it's Ligaments as well the round as broad, and all it's Veſſels had a conformation intirely like to that, which theſe ſame Parts have in Women. The Teſticles, which were ten lines long, and two broad, were as in Women, compoſed of a great number of ſmall Bladders, and faſtned near the Membranes which are at the extremity of the* Tubæ, *and which is called their Fringe.*

Drelincourt hath very little on this Subject, all he ſaith is, *Urethra rubicunda ſolida & brevis. Vagina admodum rugoſa, monticulum habens in medio, Papillis extuberans ut in Palato, Pollicem longa,tranſverſim ſciſſa, Pollicem lata. Orificium interiùs valdè ſolidum. Cervix interior admodùm dura, & paulò intrà oſculum internum duritie cartilaginoſâ.*

We ſhall proceed now to the Parts of the *Middle Venter*,the *Thorax* ; and here, as the Parts are fewer, ſo my Remarks will be alſo : and the rather, becauſe in our *Pygmie* we obſerved ſo very little difference from the Structure of the ſame Parts in a *Man.* I muſt confeſs I can't be ſo particular in all Circumſtances, as I would, becauſe for the preſerving the *Sceleton* more entire, I did not take off the *Sternum.* However, I obſerved enough to ſatisfie my ſelf with what I thought moſt material.

This *Cavity* was divided from the *Abdomen* by the *Diaphragm*, whoſe *Aponeuroſis* or *Tendon* ſeemed rather larger than in a *Man :* and the ſecond *Muſcle* which encompaſſed the *Gula*, as it paſſes through it, was very fair.

I made no Remarks upon the *Pleura*,and *Mediaſtinum :* The *Thymus* in our *Pygmie* was about an Inch long, and placed as 'tis in *Man* ; downwards 'twas divided, but upwards 'twas joyned together. So in a *Man* I have often obſerved it divided. Generally this part is larger in *Infants* and *Embrios* than in *grown Perſons*, for the Reaſons I have frequently mentioned in my *Anatomical Lectures.* The *Pariſians* obſerved in their *Monkeys* that the *Thymus* was large. *Blaſius* and *Drelincourt* have no Remarks about it.

The *Lungs* in our *Pygmie* had three *Lobes* on one ſide, and but two on the other ; five in all. Their Colour, Subſtance, Situation, and all Circumſtances exactly reſemble a *Man's.* The *Pariſians* tell us, that in their *Monkeys the Lungs had ſeven Lobes, three on the right ſide, and as many*

many on the left, the ſeventh was in the Cavity of the Mediaſtine, *as in the generality of Brutes. This again makes a notable difference between the internal parts of the* Ape, *and thoſe of* Man, *whoſe* Lungs *have generally at the moſt but* five Lobes, *oftener but* four, *and ſometimes but* two. Veſalius *affirms that he never ſaw in Man this fifth Lobe, which he reports to be in* Apes, *ſuppoſing that they have but* five. The Paſſage that the *Pariſians* hint at in *Veſalius* is this, *Lobum autem qui in Canibus, ſimiiſque Venæ Cavæ Candicem ſuffulcit, nunquam in homine obſervavi, & hunc illo deſtitui certo certius ſcio, quamvis interim* Galeni *locus in ſeptimo de adminiſtrandis Diſſectionibus mihi memoria non exciderit, quo* inquit, *quintum hunc Pulmonis Lobum eos non latère, qui recte ſectionem adminiſtrant ; innuens* Herophilo *&* Marino *ejuſmodi Lobum fuiſſe incognitum, uti ſanè fuit, cùm illi Hominum Cadavera, non autem cum ipſo, ſimiarum ac Canum duntaxat aggrederentur, in quibus præſenti Lobo nihil eſt manifeſtius.* (72) Tho' *Galen* be herein miſtaken, *Veſalius* certainly is too ſevere in his Cenſure, in charging him, that he never diſſected any thing but *Apes* and *Dogs* ; for the contrary evidently appears in abundance of Inſtances, that might be produced. And one would think he had not diſſected *Apes* and *Monkeys* in making but five *Lobes* in their *Lungs*, whereas in either there are more. In what he argues, that this *fifth Lobe* in a Man could not lie upon the *Vena Cava* ; becauſe in a *Man* the *Pericardium* is faſtened to the *Diaphragm*, and the *Vena Cava* enters there, and ſo immediately paſſes to the *Heart* ; this is true, and the ſame I obſerved in our *Pygmie.* So that in the formation of this Part, our *Pygmie* exactly reſembles a *Man* ; and is different from both the *Monkey* and *Ape*-kind. The former we have ſeen ; as to the latter, *Drelincourt* tells us in the *Male Ape* ; *Pulmo dexter quadrifidus , Lobus infimus omnium craſſiſſimus , ſuperior minùs craſſus , intermedius reapſè medius ſitu & magnitudine. Quartus demùm crenam inſculptam habet , quâ parte Cavæ fulcrum præbet. Pulmo ſiniſter bifidus , & Lobus ejus ſuperior biſurcatus.* So in the *Female* Ape, *Lobi Pulmonis dextri totalitèr diviſi IV, quorum ſuperior , bifidus totus , adeo ut ſint quinque in eâ parte : ſiniſter Pulmo bifidus totus, & Lobus ſuperior ultrà dimidium ſui diviſus.*

The *Trachæa* or *Wind-pipe* in our *Pygmie* was altogether the ſame as in a *Man* ; conſiſting of a regular order of *Cartilaginous Annuli*, which were not perfectly continued round ; but towards the *Spine*, were joyned by a ſtrong Membrane. *Drelincourt* ſaith of it, *Trachææ annuli ſe habent uti Inteſtinorum ſpiræ, nervoſis Membranis colliguntur.* The Compariſon, I think, is not ſo well made.

---

(72) *Andr. Veſalij de Corporis humani fabrica,* lib. 6. cap. 7. p. 724.

For the prefent we will leave following the Duct of the *Trachæa* up to the *Larynx*, (the Part according to the Method of *Nature*, we fhould have began with) and make fome farther Obfervations, on thofe under our prefent view. In the Cavity of the *Thorax* therefore, (as I have re-mark'd) the *Pericardium* or that *Bag* that inclofes the *Heart* in our *Pyg-mie*, was faftened to the *Diaphragm*, juft as 'tis in *Man*. I muft confefs, when I firft obferved it, I was furprifed, becaufe I had not feen it fo in *Brutes* before. And *Vefalius*, and others make it as a peculiarity to a *Man*. I will quote *Vefalius*'s words, and make an Inference from our Obfervation, and fo proceed.

Vefalius (73) therefore tells us, *Cæterum Involucri mucro, & dextri ip-fius lateris egregia portio Septi tranfverfi nerveo circulo validiſſimè, amploque admodùm ſpatio connafcitur, quod Hominibus eſt peculiare. Simiis quoque & Canibus & Porcis involucrum à ſepto multùm diſtat. Tantùm abeſt ut ipſi magnâ ſui portione connecteretur, adeò ſanè ut & hinc luce clarius conſtet, Galenum hominis viſcera aut oſcitantẏr, aut neutiquàm ſpectáſſe, Simiaſque & Canes nobis deſcribentem, immerito veteres arguiſſe.* He can't forbear at all turns to have a fling at *Galen:* But he is here in the right, and *Galen* miftaken. So *Blancardus* (74) tells us, *Homo præ cæteris Animalibus hoc peculiare habet, quod ejus Pericardium Septi tranfverfi medio ſemper accreſcat, cum idem in Quadrupedum genere liberum, & aliquanto ſpatio ab ipſo remotum ſit.*

Now our *Pygmie* having the *Pericardium* thus faftened to the *Dia-phragm*, it feems to me, as if Nature defigned it to be a *Biped* and to go *erect*. For therefore in a Man is the *Pericardium* thus faftened, that in *Expiration* it might affift the *Diaſtole* of the *Diaphragm:* for otherwife the *Liver* and *Stomach* being fo weighty, they would draw it down too much towards the *Abdomen* ; fo that upon the *relaxation* of it's Fibres in it's *Diaſtole*, it would not afcend fufficiently into the *Thorax*, fo as to caufe a Subfidence of the *Lungs* by leffening the Cavity there. In *Qua-drupeds* there is no need of this adhæfion of the *Pericardium* to the *Dia-phragm:* For in them, in *Expiration*, when the Fibres of the *Diaphragm* are relaxed, the weight of the *Vifcera* of the *Abdomen* will eafily prefs the *Diaphragm* up, into the Cavity of the *Thorax*, and fo perform that Service. Befides, was the *Pericardium* faftened to the *Diaphragm* in *Quadrupeds*, it would hinder it's *Syftole* in *Infpiration* ; or it's defcent downwards upon the contraction of it's *Mufcular Fibres* ; and the more, becaufe the *Diaphragm* being thus tied up, it could not then fo freely force down the weight of the *Vifcera*, which are always prefling upon it, and confequently not fufficiently dilate the Cavity of the *Thorax* , and therefore muft hinder their *Infpiration*. Thus we fee how neceffary it is,

(73) *Andr. Vefalij de Corporis Humani fabrica,*lib.6.cap.8.p.m.728. (74) *Steph: Blancardi Anatom. reformat.* cap.2. p.8.

H that

that in a *Man* the *Pericardium* ſhould be faſtened to the *Diaphragm* ; and in *Quadrupeds* how inconvenient it would be ; that from hence I think we may ſafely conclude, that *Nature* deſign'd our *Pygmie* to go erect, ſince in this particular 'tis ſo like a *Man* ; which the common *Apes* and *Monkeys* are not ; and tho' they are taught to go erect, yet 'tis no more than what *Dogs* may be taught to do.

We proceed now to the *Heart* ; where we obſerved that in our *Pygmie*, it's *Auricles*, *Ventricles*, *Valves* and *Veſſels* were much the ſame as they are in a *Man's*. It's *Cone* was not ſo pointed, as in ſome *Animals*, but rather more obtuſe and blunt, even more than a *Man's*. What *Avicenna* (75) remarks of the *Heart* of an *Ape*, having a *double Cone*, muſt be accidental and extraordinary : for he tells us, *Et jam repertum eſt Cor cujuſdam Simij habens duo Capita*. And a little after, he denies the *Heart* to be a *Muſcle* ; *Jam autem erravit* (ſaith he) *qui exiſtimavit, quòd ſit Lacertus, quamvis ſit ſimilium rerum in eo, verùm motus ejus non eſt voluntarius*. The Perſon he hints at, I ſuppoſe, is *Hippocrates*, who ſo long ago aſſerted this ; Ἡ καρδίη (ſaith (76) *Hippocrates*) μῦς ὅχιν καρδία ἰχυρὸς, ὃ τῷ νεύρῳ, ἀλλὰ πληματι σαρκός. *Cor muſculus eſt validus admodùm non Nervo, verùm Carnis ſpiſſamento*. And *Steno* and Dr. *Lower* ſince have ſhewed us the way of diſſecting it, and have made it moſt evident that 'tis Muſcular ; and it's *motion* is ſuch ; but as *Avicenne* obſerves , 'tis not a *voluntary* motion, but *involuntary*. 'Tis pity we had not a better *Tranſlation* of *his Works* ; for unleſs it be ſome particular Pieces, the reſt is moſt barbarouſly done, as appears from that little I have quoted of him. But to return to our *Pygmie* ; the magnitude and figure of the *Heart* here, was exactly the ſame as repreſented in our *Scheme*, where part of the *Pericardium* is left lying on it. Both in the right and left *Auricle* and *Ventricle*, I obſerved two *Polypous Concretions*, which plainly repreſented the *Valves* both in the *Arteria Pulmonalis*, and *Aorta*. I muſt confeſs by what I have hitherto obſerved of them, (and I have very frequently met with ſuch *Concretions* in *Humane Bodies*) I cannot think theſe *Polypus*'s to be any thing elſe, than the *Size* of the Blood, or the *Serum coagulated* after Death. The Obſervation I formerly gave (77) of a *Polypus* in the *Trachæa* and *Bronchiæ* of a Patient troubled with an *Hæmoptoe*, in it's kind I think remarkable.

The *Pariſians* obſerve that *the Heart of their* Monkey *was a great deal more pointed, than it uſually is in* Man ; *which is likewiſe a Character of* Brutes. *Yet in the Interiour* Superficies *of it's* Ventricles , *it had that great number of Fibres and* fleſhy Columns, *which are ſeen in* Men. *Drelincourt* in his *Ape* obſerves, *Cor ſolidum in ventriculo ſiniſtro, laxum in dextro* ; *prædurus Conus ejus : Serum in Pericardio ſalſum. Vaſa Coronaria tumida, præſertim circà* Ventriculum. *Adeps circà ea nullus.*

(75) *Avicenna* lib. 3. Fen. 2. Tract. 1. p.m.670   (76) *Hipp. de Corae*, p.m.270.   (77) *Vide Th. Bartholini Acta Med. & Philoſ. Haſnienſ. Vol.* 5. *Obſ.* 30. p.94.

There

There was nothing farther, I think, that I obferved peculiar in the *Thorax* of our *Pygmie*. I fhall now therefore follow the Duct of the *Trachæa* up to the *Throat*. And here as in *Man*, I obferved placed the *Glandula Thyroidca*, upon the *Cartilago Scutiformis* of the *Larynx* ; 'twas red and fpungy, full of Blood veffels, not much unlike the inward Part of the *Spleen*, but fomewhat firmer. In a Man I have always obferved this part to be red. *Drelincourt*'s Account of it in the *Ape* is , *Glandulæ Thyroideæ & Cricoideæ craffæ funt, & fubnigricantes ; & illas permeant furculi Corotidis Arteriæ & Jugularis venæ externæ ; cum furculis Nervi Recurrentis.* There is no fenfible account yet given of the ufe of this part, as I have met with : And I think that from a Comparative Survey of it in other Animals, and a ftrict Obfervation of it's Structure, and the Veffels that compound it , it were not difficult to affign other ufes of it more fatisfactory.

As to the *Larynx* in our *Pygmie*, unlefs I enumerate all the *Cartilages* that go to form it, and the *Mufcles* that ferve to give them their Motion, and the *Veffels* that run to and from it, and the *Membranes* and *Glands* , there is nothing that I can further add, but only fay, that I found the whole Structure of this Part exactly as 'tis in *Man*. And the fame too I muft fay of the *Os Hyoides*. The *Reflection* that the *Parifians* make upon the obfervation of this, and it's neighbouring Parts in the Diffection of their *Monkey*'s, I think is very juft and valuable. And if there was any farther advantage for the forming of *Speech*, I can't but think our *Pygmie* had it. But upon the beft Enquiry, I was never informed, that it attempted any thing that way. Tho' *Birds* have been taught to imitate *Humane Voice*, and to pronounce Words and Sentences, yet *Quadrupeds* never ; neither has this *Quadru-manous Species* of *Animals*, that fo nearly approaches the Structure of *Mankind*, abating the *Romances* of *Antiquity* concerning them.

The *Parifians* therefore tell us , *That the Mufcles of the Os Hyoides,* Tongue, Larynx, *and* Pharynx, *which do moft ferve to articulate a word, were wholly like to thofe of* Man ; *and a great deal more than thofe of the* Hand ; *which neverthelefs the* Ape, *which fpeaks not, ufes almoft with as much perfection as a* Man. *Which demonftrates, that Speech is an Action more peculiar to* Man, *and which more diftinguifhes him from Brutes than the* Hand ; *which* Anaxagoras, Ariftotle, *and* Galen *have thought to be the* Organ *which Nature has given to* Man, *as to the wifeft of all Animals* ; for want perhaps of this Reflection : For the Ape is found provided by Nature of all thofe marvellous Organs of Speech with fo much exactnefs, that the very three fmall Mufcles, which do take their rife from the Apophyfes Styloides, are not wanting, altho' this Apophyfis be extreamly fmall. This particularity does likewife fhew, that there is no reafon to think, that Agents do perform fuch and fuch actions, becaufe they are found with Organs proper there-

H 2   *nete*

*unto ; for, according to theſe Philoſophers, Apes ſhould ſpeak, ſeeing that they
have the Inſtruments neceſſary for Speech.*

I ſhall not engage in this Argument here, becauſe it would be too
great a digreſſion ; hereafter, it may be, I may take an occaſion to do it ;
for this is not the only Inſtance in our Subject, that will juſtifie ſuch an
Inference : tho' I think it ſo ſtrong an one, as the *Atheiſts* can never anſwer.

We ſhall take notice next of the *Uvula,* a Part of ſome uſe too in
forming the *Voice* ; for where 'tis miſſing or vitiated, it much alters the
ſound ; and even this I found in our *Pygmie* to be altogether alike as in
Man. It had thoſe two Muſcles which are in a *Man,* the *Muſculus Sphæ-
no-Palatinus,* and the *Pterigo-Palatinus ſeu Sphæno-Pterigo-Palatinus* ; the
Tendon of which laſt, paſſed over the *Pterigoidal Proceſs,* which was to
it like a *Trochlea* or *Pully,* and was afterwards inſerted as in a *Man.*

The *Pariſians* tell us that *the* Uvula, *which is in no other Brutes, was
found in our* Apes (it ſhould be *Monkeys*) *wholly reſembling that of Man.*
And ſo *Blaſius, Uvula in Animalibus aliis præter hominem & ſimiam nunquam à me obſervata.* All that *Drelincourt* ſaith of it is, *Uvula firma
eſt & carnoſa.*

The *Tongue* of our *Pygmie* in all reſpects, as I know of, reſembled a
Humane Tongue ; only becauſe 'twas ſomewhat narrower, it ſeemed longer : And under the *Tongue* in our *Pygmie* we obſerved the *Glandulæ Sub-
linguales* as in *Man.*

Drelincourt obſerves in the *Ape, Linguæ baſis non tantùm incumbit Hy-
oidi ſuperno, ſed amplectitur ejus tuber inferius poſticè : Papillas habet Bovinis
ſimiles, & tunicam propriam permeantes.*

At the Root of the *Tongue* of each ſide were placed the *Tonſillæ* in our
Pygmie, as they are in a *Man.* They were protuberant and hard, and
not ſo foraminulous, as uſually in Man ; very probably being vitiated
by the *Ulcer* in the Cheek. For *Drelincourt* tells us in the *Ape, Amygdalæ
cavæ, pertuſæ & Scrobiculos habentes.*

The *Parotides* under each *Ear* in our *Pygmie* were large, and of the
ſame Figure as in *Man.* *Parotis glandula contegit Muſculum Sterno-Maſtoi-
deum, articulationem Maxillæ & Muſculi Pectoralis portionem,* ſaith *Drelincourt.*

The *Maxillary Gland* of the left ſide (where the Ulcer in our *Pygmie*
was) had two of it's Lobes, globous and protuberant, above the Surface
of the other Part, being infected and tumefied by the Ulcerous Matter.
Theſe *Glands* were about an Inch long, and about half an Inch broad ;
and

and there were two other fmall *Glands* a little diftant from the head of the *Maxillary. Glandulæ falivales ad angulum Maxillæ Inferioris oblongæ, laxæ, molles, albicantes,* faith *Drelincourt.*

But before I leave thefe Parts, there are fome others I muft here take notice of, in this *Comparative Survey;* which tho' they are not to be met with, either in our *Pygmie* or in *Man;* yet are very remarkable, both in the *Monkey* and *Ape-kind,* viz. thofe *Pouches* the *Monkeys* and *Apes* have in their *Chaps,* which ferve them as *Repofitories* for to hoard up, upon occafion, food in; when they are not difpofed for the prefent to devour it; but when there *Stomachs* ferve them, they then take it out thence and fo eat it. That the *Ape-kind* has thefe *Pouches, Drelincourt* does inform us; where he tells us, *Mufculus latiffimus Mentum univerfum & buccas obtegit, quâ parte fimiæ faccum formant, intra quem Efculenta recondunt. Pliny* is very exprefs, That both *Satyrs* and *Sphinges* (which I make to be of the *Monkey-kind*) have them likewife, (78) *Condit in Thefauros Maxillarum Cibum Sphingiorum & Satyrorum Genus. Mox inde fenfim ad mandendum manibus expromit; & quod formicis in annum folenne eft, his in dies vel horas.* The account the *Parifians* give us of this *Pouch* in the Mouth of the *Monkeys* they diffected, is this; *That it was compofed of Membranes and Glands, and of a great many Mufculous and Carnous Fibres. It's fituation was on the outfide of each Jaw, reaching obliquely from the middle of the Jaw to the under part of it's Angle, paffing under a part of the Mufcle called* Longiffimus. *It was an Inch and an half long, and almoft as broad towards it's bottom. It opened into the Mouth between the Jaw and the bottom of the Gum.* 'Tis into *this Pouch that Apes ufe to put what they would keep; and it is probable that the Mufculous Fibres which it has, do ferve to fhut and open it, to receive and put out what thefe Animals do there lay up in referve.* Now our *Pygmie* having none of thefe *Pouches* in it's *Chaps,* nor nothing like them; 'tis a notable difference both from the *Monkey* and *Ape-kind,* and an Agreement with the *Humane.*

We fhould now come to difcourfe of the *five Senfes:* But there is little I have at prefent to remark of them. For in the *Organs* of thofe of *Tactus,* and *Guftus,* there was no difference I could obferve between our *Pygmie* and a *Man.* As to thofe of *Hearing* and *Smelling,* I fhall make my Obfervations upon them, in the *Ofteology.* Here therefore I fhall only remark fome things of the *Eyes,* the *Organs* of Seeing; and fo proceed to the *Brain.*

The Bony *Orbit* of the *Eye* in our *Pygmie* was large, conical, and deep. Here we obferved the *Glandula Lachrymalis,* and *Innominata.* The *Bulb* of the *Eye* in proportion to the Bulk of the Body, was rather larger than in a *Man.* The *Iris* was of a light hazel Colour: The *Pu-*

(78) *Plinij Hift. Nat.* lib. x. cap. 72. p. m. 466.

pil.

pil round and large : The *Cryftalline Humour* Sphærical or *Lentiformis*,and almoft as large as in a *Man*.   The *Optic Nerve* was inferted exactly as in a *Man*.   The *Tunica Choroides* rather blacker than in a *Man*.   And whereas in *Brutes*, that are *prono Capite* , there is ufually a *Mufculus Septimus*, which from it's ufe is call'd *Sufpenforius* ; in our *Pygmie* there was none of this *Mufcle*.   All the other *Mufcles* of the Eye, were exactly the fame as in *Man*.   This *feventh Mufcle* is alfo wanting in the *Ape*, as appears by the figures *Cafferius* (79) has given us of the *Eye* of an *Ape*.   Neither the *Parifians*, nor *Blafius*, nor *Drelincourt* do give us any Remarks upon this *Part*.

We proceed now to the *upper Venter*, the *Head*, where at prefent we fhall examine the *Brain* ; that Part, which if we had proceeded according to the *Method of Nature* in forming the Parts, we muft have began with.   For I can't but think, as 'tis the firft Part we obferve formed, fo that the whole of the Body, *i. e.* all the *Containing Parts*, have their rife from it.   But I fhall not enlarge upon this Argument here ; it would be too great a digreffion, to give my Reafons for fuch an *Hypothefis*. From what is generally received, *viz.* That the *Brain* is reputed the more immediate Seat of the *Soul* it felf ; one would be apt to think, that fince there is fo great a difparity between the *Soul* of a *Man*, and a *Brute*, the *Organ* likewife in which 'tis placed fhould be very different too. Yet by comparing the *Brain* of our *Pygmie* with that of a *Man* ; and, with the greateft exactnefs, obferving each Part in both ; it was very furprifing to me to find fo great a refemblance of the one to the other, that nothing could be more.   So that when I am defcribing the *Brain* of our *Pygmie*, you may juftly fufpect I am defcribing that of a Man , or may think that I might very well omit it wholly, by referring you to the accounts already given of the *Anatomy* of an *Humane Brain*, for that will indifferently ferve for our *Pygmie*, by allowing only for the magnitude of the Parts in *Man*.   Tho' at the fame time I muft obferve , that proportionably to the Bulk of the Body, the *Brain* in our *Pygmie*, was extreamly large ; for it weighed (the greateft part of the *Dura Mater* being taken off) twelve Ounces, wanting only a Dram.   The *Parifians* remark, that in their *Monkeys the Brain was large in proportion to the Body, it weighing two Ounces and a half :* which neverthelefs was inconfiderable to ours ; fince our *Pygmie* exceeded not the Stature and Bulk of the Common *Monkey* or *Ape* ; fo that herein, as in a great many other Circumftances, our *Pygmie* is different from the Common *Monkey* and *Ape* , and more refembles a *Man*.

I can't agree with *Vefalius*, that the Structure of the *Brain* of all *Quadrupeds*, nay all *Birds*, and of fome *Fifhes* too, is the fame as in Man.

(79) *Jul. Cafferij Placentini Pentaftthefeion*, h. e. de quinque fenfibus. *Vide de Organo vifus*. Tab. 3. fig. 12, 13.

There

There is a vaſt difference to be obſerved in the formation of the Parts, that ſerve to compoſe the *Brain* in theſe various *Animals.* And tho' the *Brain* of a *Man*, in reſpect of his Body, be much larger than what is to be met with in any other *Animal* (for *Veſalius* makes the *Brain* of a *Man* to be as big as thoſe of three Oxen) yet I think we can't ſafely conclude with him, that *Animals*, as they excell in the largeneſs of the *Brain*, ſo they do likewiſe in the Principal Faculties of the *Soul :* For if this be true, then our *Pygmie* muſt equal a *Man*, or come very near him, ſince his *Brain* in proportion to his Body, was as large as a *Man's.* *Veſalius* (80) his words are theſe ; *Cerebri nimirùm conſtructione Simia, Canis, E- quus, Felis, & Quadrupeda quæ hactenùs vidi omnia, & Aves etiam uni- verſæ, plurimaque Piſcium genera, omni propemodùm ex parte Homini cor- reſpondeant : neque ullum ſecanti occurrat diſcrimen, quod ſecus de Hominis, quàm illorum Animalium functionibus ſtatuendum eſſe præſcribat. Niſi ſeriè quis meritò dicat Cerebri molem Homini, Perfectiſſimo ſanè quod novimus Animali, obtigiſſe maximam, ejuſque Cerebrum etiam tribus Boum Cerebris grandius reperiri : & dein ſecundum Corporis proportionem mox Simiæ, dein Cani magnum quoque non ſecus obtingere Cerebrum, quàm ſi Animalia Cerebri tantum præſtarent mole, quanto Principis Animæ viribus apertius viciniùſve donata videntur.*

Since therefore in all reſpects the *Brain* of our *Pygmie* does ſo exactly reſemble a *Man's*, I might here make the ſame Reflection the *Pariſians* did upon the *Organs of Speech*, That there *is no reaſon to think*, that A- gents do perform ſuch and ſuch Actions, becauſe they are found with Organs proper thereunto : for then our *Pygmie* might be really a *Man.* The *Or- gans* in *Animal* Bodies are only a regular *Compages* of Pipes and Veſſels, for the *Fluids* to paſs through, and are paſſive. What actuates them, are the *Humours* and *Fluids :* and *Animal Life* conſiſts in their due and regu- lar motion in this *Organical* Body. But thoſe *Nobler Faculties* in the *Mind of Man*, muſt certainly have a *higher Principle ;* and *Matter orga- nized* could never produce them ; for why elſe, where the *Organ* is the ſame, ſhould not the *Actions* be the ſame too? and if all depended on the *Organ*, not only our *Pygmie*, but other *Brutes* likewiſe, would be too near akin to us. This *Difference* I cannot but remark, that the *Ancients* were fond of making *Brutes* to be *Men :* on the contrary now, moſt un- philoſophically, the *Humour* is, to make *Men* but meer *Brutes* and *Mat- ter.* Whereas in truth *Man* is part a *Brute*, part an *Angel ;* and is that *Link* in the *Creation*, that joyns them both together.

This *Digreſſion* may be the more pardonable, becauſe I have ſo little to ſay here, beſides juſt naming the *Parts ;* and to tell you (what I have alrea- dy) that they were all like to thoſe in a *Man :* For the *Dura Mater*, as a *Common Membrane*, firmly ſecured the ſituation of the whole *Brain*, ſtrictly

(80) *And. Veſalij de Corporis Humanis fabrica,* lib. 7. cap. 1. p. 773, 774.

adhering :

adhering to the *Sutures* of the *Cranium* above ; before to the *Crista Galli* ; and at the *basis* so strongly, that it was not easily to be separated. By it's *anterior Procefs* of the *Falx*, it divided the two *Hemispheres* of the *Cerebrum*; by it's *transverse Procefs*, which descended deep, just as in a *Man* , it separated the *Cerebrum* and *Cerebellum :* it enjoyed the same *Sinus*'s, and in all Particulars 'twas conformable to what is in a *Man*.

The *Pia Mater* in our *Pygmie* was a fine thin Membrane which more immediately covered the Substance of the *Brain*, and may be reckoned it's *proper Membrane* ; infinuating it's felf all along between the *Anfractus* of the *Cerebrum* and the *Circilli* of the *Cerebellum* ; being copiously furnished with numerous Branches of *Blood Veffels*, but they appear'd more on the *Convex* Part, then at the *Bafis*.

The *figure* of the whole *Brain* in our *Pygmie* was globous ; but by means of a greater jutting in of the Bones of the *Orbit* of the *Eye*, there was occasioned a deeper depreffion on the *Anterior Lobes* of the *Brain* in this place, than in a *Man*. As to other Circumftances here, I obferved all Parts the fame. The *Anfractus* of the *Cerebrum* were alike ; as alfo the *Subftantia Corticalis* and *Medullaris*. On the *basis* of the *Brain*, we may view all the *Ten pair* of *Nerves* exactly situated and placed as in a *Humane Brain* ; nor did I find their Originations different, or any Particularity that was fo. I fhall therefore refer to the *figures* I have caufed to be made of the *Brain*, and their Defcriptions ; where we may obferve the *Arteriæ Carotides*, *Vertebrales*, and *Communicans*, and the whole of the *Blood Veffels* in our *Pygmie* to be the fame as in a Man. Here was the *Infundibulum*, the *Glandulæ duæ albæ ponè Infundibulum*, the *Medulla Oblongata* with it's *Annular Protuberance*, and the beginning of the *Medulla Spinalis*, juft as in *Man*. I am here only a *Nomenclator*, for want of Matter to make particular Remarks upon. And the Authorrs that have hitherto furnished me with Notes, how the fame *Parts* are in *Apes* and *Monkeys*, do fail me now ; it may be, finding here nothing new or different, they are therefore filent. All the *Parifians* do tell us of the *Brain* in their *Monkeys* is this :

The Brain *was large in proportion to the Body : It weighed two Ounces and an half. The* Dura Mater *entered very far to form the* Falx. *The Anfractuofities of the External part of the Brain were very like thofe of Man in the Anteriour part ; but in the inward parts before the Cerebellum, there was hardly any : they in requital were much deeper in proportion. The* Apophyfes, *which are called* Mamillares, *which are great Nerves that do ferve to the fmelling, were not foft, as in Man, but hard and membranous. The* Optick Nerves *were alfo of a Subftance harder and firmer than ordinary. The* Glandula Pinealis *was of a Conical figure, and it's point was turned towards the hinder part of the Head. There was no* Rete Mirabile : *for the* Carotides *being entered into the Brain, went by one fingle Trunk on each fide*
                                                                                    of

of the edge of the feat of the Sphenoides *to pierce the* Dura Mater, *and to be diftributed into the bafis of the Brain.* In our Subject I thought the *An-fractus* of the Brain much the fame, both in the anteriour and hinder part. Nor did I obferve any difference in the *Mamillary Proceffes* or *Op-tick Nerves*, or *Rete Mirabile*, but all, as in a *Man*.

The *Cerebellum* in our *Pygmie* was divided by *Circilli*, as in *Man*. It had likewife the *Proceffus Vermiformes*. Dr. *Willis* (80) makes this Re-mark upon this Part : *Cerebellum autem ipfum, in quibufvis ferè Animali-bus, ejufdem figuræ & proportionis, nec non ex ejufmodi lamellis conflatum reperitur. Quæ Cerebrum diverfimodè ab homine configuratum habent , ut Volucres & Pifces, item inter Quadrupedes* Cuniculi & Mures, *quorum Ce-rebra gyris feu convolutionibus carent ; his Cerebelli fpecies eadem, fimilis pli-carum difpofitio & Partium cæterarum compofituræ exiftunt.* 'Tis from hence he forms his noted *Hypothefis*, How that the *Animal Spirits* that are bred in the *Cerebrum* , do ferve for *Voluntary Motions* ; and thofe in the *Cerebellum* for *involuntary*.

If we furvey the inward Parts of the *Brain* in our *Pygmie*, we fhall here likewife find all exactly as in a *Humane Body* ; viz. The *Medullary* Subftance running up between the *Cortical* ; The *Concameration*, the *Cor-pus Callofum*, the *Fornix* and it's *Crura* the fame. The *Ventricles* large and fpatious. The *Corpora Striata*, the *Thalami Nervorum Opticorum* all alike. The *Plexus Choroides* the fame ; as were alfo the three *Foramina* as in *Man* ; The *Glandula Pinealis* proportionably large. The *Protuberantiæ Orbiculares* ; i. e. The *Nates* and *Teftes* in our *Pygmie* were the fame as in *Man* ; whereas in *Brutes* (as Dr. *Willis* well obferves) the *Nates* are al-ways proportionably larger than in *Man* ; but it was not fo in our *Pyg-mie*. The *Valvula major* here was very plain. The *Cerebellum* being di-vided, the *Medullary* Parts reprefented the Branches of Trees, as a *Man's* does. The *Medulla Oblongata* and *Medulla Spinalis* the fame as the *Hu-mane* ; and all Parts being fo conformable here to a *Humane Brain*, I thought it fufficient juft to name them, fince I have caufed to be made two *figures* of the *Brain* in our *Pygmie* from the Life, and in its Natural Bignefs, where all the Parts are plainly reprefented to the Eye.

---

(80) *Willis Cerebri Anat.* cap.3. p.22.

I   T H E

·THE

OSTEOLOGY,

O R

DESCRIPTION

OF THE

BONES.

WE come now to the *Ofteology*, to give a Defcription of
the *Sceleton* of our *Pygmie*, by comparing which, with
that of a *Man*, an *Ape* and a *Monkey*, we may obferve
(as we have already of the other Parts) that here too,
our *Pygmie* more refembles a *Man* than *Apes* and *Monkey's* do ; but where
it differs, there 'tis like the *Ape-kind*. *Galen* (as I have already quoted
him, *vid. p.* 15.) tells us that *an Ape of all Creatures is the moft like to a
Man in the* Vifcera, *Mufcles, Arteries, Veins and Nerves, becaufe 'tis fo
in the Structure of the Bones.* But it may be queftioned , Whether even
the Structure of the *Bones* themfelves, does not depend upon that of the
Mufcles : fince in their firft Formation, they are *foft* and *vafcular* ; then
Cartilaginous, and in time at laft are hardened into *Bones.* In *Ricketty
Children* too, we find, that even the *Bones* are rendered crooked, by the
Contraction of the *Mufcles*, how much more, when they are tender and
foft, might they be bended any way by them. But by underftanding
exactly the Structure of the *Bones*, we fhall the better apprehend the
Rife and Infertions of the *Mufcles*. And for the better attaining this ,
Galen in the fame Chaper (81) recommends to his *Students*, when they
cannot have an opportunity of Confulting an *Humane Sceleton* , then to

(81) *Galen. de Anat. adminiftr.* lib. 1. cap. 3. p. m. 29, 30..

make

make use of those of *Apes* ; not that he thinks them both alike, but the most like : and tells them, that it was worth their while on this account to go to *Alexandria*, where the Phyficians taught their Scholars the Doctrine of the Bones from the Infpection of *Humane Sceletons* themfelves, which he much prefers before Books. But fince in his time *Humane Sceletons* were not to be had but at *Alexandria*, for the fupplying this Defect, they might obferve the *Bones* of *Apes* ; and after that, they might read his Book *De Offium Naturâ*, and to do as he did, vifit the *Sepulchres* and Graves, and to obferve there the *Humane Bones* themfelves: And he tells us of two *Sceletons* he made ufe of ; One that the River had wafhed out of a *Tomb*, where the Flefh was corrupted and wafhed away, yet the Bones held together. The other was of a Thief that was Executed, who was fo much hated, that none would bury him ; but the Birds pick'd off his Flefh, and left his *Bones* as a *Sceleton*. But faith he, *σὺ ᾖ εἰ μηδὲ τοιετον μηδὲν ἐυτύχησας θεάσαϑαι. πίθηκον ᾖν ἀνατεμών, ἐπ᾽ αὐτῶ καλανόησον ἕκασον τῶν ὀςῶν ἀκρεϐῶς*, &c. i. e. *If you can't happen to fee any of thefe, diffect an Ape, carefully view each Bone* , &c. Then he advifes what fort of Apes to make choice of, as moft refembling a Man : And concludes, *ἁπάντων, ὡς ἐξίω, τῶν ὀςῶν καλανόησαι χρὴ τίω φύσιν εἴτ᾽ ὅϖ ἀνθρώπε σώμαϊι, εἴτ᾽ ἀν πιθήκε εἰ δυνηθείης· ἄμεινον δ᾽ ὅϖ ἀμφοῖν, εἶθ᾽ ἑξῆς ὅϖ τὼ τῶν μυῶν ἀναϊομίω ἐλθεῖν*, i. e. *One ought to know the Structure of all the Bones either in a Humane Body, or in an Ape's ; 'tis beſt in both ; and then to go to the Anatomy of the Mufcles.*

What *Galen* advifed, no doubt he practifed himfelf , and obferved both. But *Andreas Vefalius* will not allow him this : For in his great and excellent Book *De Corporis Humani Fabricâ*, he all along tell's us, that *Galen* gives us rather the *Anatomy* of Apes than of a *Man :* And in his *Epiftola ad Joachimum Roelants de Radice Chynæ*, his chief Defign is to prove, that *Galen* never diffected a *Humane Body :* and that he is often miftaken in the Hiftory of the Parts, as alfo in their Ufes ; and that his Reafonings are frequently unconclufive.

Upon the coming out of *Vefalius* his firft Book, he was warmly oppofed by *Jacobus Sylvius* a Phyfician at *Paris*, who had formerly been *Vefalius* his Mafter in *Anatomy* ; in a Treatife ftiled *Depulfio Vefani cujufdam Calumniarum in Hippocratis & Galeni Rem Anatomicam*. This was anfwered not long after by *Renatus Henerus*, who publifhed another Treatife, *viz. Adverfus Jacobi Sylvij Depulfionum Anatomicarum Calumnias pro Andreâ Vefalio Apologia*. *Sylvius* afterwards procures a Difciple of his to write againft *Vefalius*, who puts out, but unfuccefsfully , *Apologia pro Galeno contra Andream Vefalium Bruxellenfem, Francifco Puteo Medico Vercellenfi Authore*. A Scholar of *Vefalius*, *Gabriel Cunæus*, makes a Reply to *Puteus* in his *Apologiæ Francifci Putei pro Galeno in Anatome examen*. Upon *Vefalius* his leaving *Rome*, a Difciple of his, *Realdus Columbus*, grew very famous for *Anatomy*, but ungrateful to his Mafter, as *Vefalius* com-

I 2

complains in his Book *De Radice Chynæ*, and his *Examen Obſervationum Falloppij*. But *Gabriel Falloppius* was always kinder to him, and mentions him with the greateſt Honour, and calls him *Divine*; tho' in ſeveral things he diſſents from him, which occaſioned *Veſalius* his putting out his *Obſervationum Falloppij Examen*.

Realdus Columbus was ſucceeded at *Rome* by *Bartholomæus Euſtachius*; a Man very knowing and curious in *Anatomy*, but extreamly devoted to *Galen*, as one may ſee by this Paſſage; *Ut uno verbo me expediam, talem eum eſſe (ſc. Galenum) aſſeverem, qualem opinor neminem in poſterum futurum, fuiſſe nunquam plane confirmo. Quare dubiis in rebus diſſentire ab eo honeſt? non poſſumus, ſed magis expedire, deceveque putandum eſt, illo Duce errare, quàm his illiſve Magiſtris hodie crudiri, ne dicam cum iis vera ſentire* (82). Too great a Partiality for ſo ingenious a Man. And it may be, this was one Reaſon why *Veſalius* ſo much endeavoured to leſſen *Galen*'s Authority; becauſe the Humour of the Age was ſuch, that nothing then was to be received, but what was to be met with in him. But certainly they are in the wrong, who, becauſe *Galen* is miſtaken in ſome things, do now wholly reject him, and lay him aſide as good for nothing. The wiſeſt and moſt experienced in the *Art* may read his Works, and in reading him, if juſt and not prejudiced, will acknowledge, a Satisfaction and an Advantage they have received from him.

The Deſign of *Bartholomæus Euſtachius* in writing that Diſcourſe of his, *Oſſium Examen*, is to juſtifie *Galen*, that he did not only diſſect *Apes*, but *Humane Bodies* likewiſe; and that his Deſcriptions are conformable to the Parts in *Man*, and not to *Apes* and *Monkeys*. He therefore draws a Compariſon between the *Sceleton* of an *Ape* and a *Man*; and ſhews wherein they differ; and how far *Galen*'s Deſcriptions of thoſe Parts are different from thoſe in an *Ape*. *Volcherus Coiter* has likewiſe made the ſame *Comparative Survey*, in his *Analogia Oſſium Humanorum, Simiæ & veræ & caudatæ, quæ Cynocephali ſimilis eſt, atque Vulpis*. In moſt things I find *Coiter* to follow *Euſtachius*, but *Euſtachius* I think is to be preferred, becauſe in his *Annotationes de Oſſibus*, he quotes the Texts of *Galen* at large. *Johannes Riolanus* the *Son* hath wrote upon the ſame Argument likewiſe; *viz. Simiæ Oſteologia ſive Oſſium Simiæ & Hominis Comparatio*; and he being later than either of the former, and having made uſe of all before him, he may be thought to be the moſt exact.

In giving therefore an Account of the *Oſteology* of our *Pygmie*, and the better to compare it's *Sceleton* with that of a *Man*, and an *Ape*, and a *Monkey*, I thought I could not do better, than to inſert this Diſcourſe of *Riolanus*; and by *Commenting* upon it, to ſhew wherein our *Pygmie* a-

(82) *Barthol. Euſtachij Oſſium Examen*, p. m. 189.

grees.

grees or differs. This I thought the moſt compendious way, and what other Obſervations I have, that conveniently I can't inſert in my *Comment*, I ſhall add at the cloſe of this Diſcourſe. And tho' I may be cenſured by ſome for diſcourſing ſo largely upon an *Ape*, yet this *Apology* I have to make, That 'tis an Argument that has exerciſed the Pens of the greateſt *Anatomiſts* we have had ; and ours being one of a higher degree than the Common ſort, and in ſo many Particulars nearer approaching the Structure of *Man*, than any of the *Apekind*, and it being ſo rare and uncommon, it may the more excuſe me, if I endeavour to be as particular as I can. But in ſome meaſure to avoid this Fault, I ſhall omit *Riolan*'s *firſt Chapter*, which is but Præfatory, and begin with *the ſecond*.

C A P. II.

De Capitis & Faciei Oſſibus.

Simiæ Caput (a) *rotundum eſt, humano ſimile, cynocephali verò caput oblongius. Utriuſque* (b) *Suturæ adeo ſunt obſcuræ, ut earum nullum appareat veſtigium. Propterea potiùs harmoniæ dici merentur, quam ſuturæ,quia rerum conſutarum figuram non æmulantur. Attamen* Volcherus Coiter *ſuturas attribuit ſimiis, parum ab humanis diſcrepantes. In cercopitheco ſquammiformes deſiderantur.* (c) Frontis Os *in calvariæ baſi ſede, ad conjunctionem Oſſis ſphenoidis, tranſverſâ potiùs lineâ quàm ſutura diſtinguitur : ampla oblongaque ſciſſura homines diviſum obtinent, in quam aliud Os inſtar cribri perforatum conjicitur, arctiſſimeque conſtringitur. At* (d) Simiæ Os Frontale *ea in parte omnino continuum exiſtit, & quâ naſus principium ſumit, non longè ab ea ſede, quæ frontem conſtituit, alto & rotundo foramine parumper à lateribus compreſſo, illo foramine quod nervum viſorium emittit, nonnihil ampliori, excavatum eſt. In ejus humiliori profundiorique ſede, quatuor aut quinque alia foramina recta & lata cernuntur. In ſimia caudatâ* (e) Os Ethmoides *admodum profundè in nares deſcendit, paulò infra eam regionem ex qua naſus exoritur. Harmoniâ per medium dividitur,& utrimque ab Oſſe frontis, quod etiam profundè deſcendit effingi videtur.* (f) Superficies ſellæ Sphenoidis *ad narium principium in Oſſe frontis non eſt plana & æqualis ut in homine, ſed eſt eminentiſſima. In poſteriori ſellæ eminentia glandulam excipiente, reperitur foramen exſculptum. In ſellæ* (g) *hujus ſubſtantia nulla latet cavitas ut homine.* (h) Cavitates *illæ quæ in apophyſibus pterygoideis exſculptæ ſunt, maximæ & profundæ apparent.* (i) Oſſa, Bregmatis, & Temporum, *nec intus, nec foris, ullam demonſtrant diviſionem, quaſi ex unico continuatoque Oſſe conſtarent.* (k) *In Oſſe temporum apophyſis* Maſtoidis *deeſt,* Styloidis *exigua eſt.* (l) *Cavitas auris videtur unica, orbiculatim in plures gyros ſtriata, nec tria Oſſicula* Malleolus, Incus, & Stapes *reperiuntur, quibus aliorum animantium aures inſtructæ ſunt, ſi credimus* Caſſerio, *ſed ego ſemper obſervavi.* O₃

Os Zygœma (m), *quâ parte ab Oſſe orbitario procedit, craſſum & robuſtum eſt, atque ejus in medium lineâ potiùs quàm ſuturâ diſtinguitur. In homine verò tenue exiſtit, & ſuturâ dirimitur.*

Simiæ Facies (n) *rotunda eſt, cynocephali oblonga & antrorſum protuberans. Oſſa verò* Maxillæ ſuperioris *reſpondent humanis.*. (o) *Suturæ ſunt harmoniis, ſive rimis ſimiles, potiſſimum ea quæ medium palatum interſecat. Sed peculiaris ſutura notatur, ab inferiore orbita incipiens, ſecundum longitudinem maxillæ ad caninum dentem cujuſque lateris prorepit, ipſumque palatum dirimit.*

Maxilla inferior (p) *integra eſt, nullâ lineâ in mento diſſecta, breviſſima eſt corporis proportione, ita ut ex omni animantium genere nullum breviorem habeat, excepto homine. Extremitas quæ, cavitati temporum articulatur, eſt condyloïdis, ut in homine. Quare non eſt gynglymoides hæc articulatio, ut ſcripſit* Volcherus Coïter.

(q) *Simia in* dentibus *caninis & molaribus differt ab homine.* · Caninos *quidem habet* dentes *humanis ſimiles, in unaquaque maxilla binos, qui utrimque aſſident & accumbunt inciſoribus. Singulas radices ut inciſores habent, ſed altiùs infixas robuſtioreſque, minùs etiam ex anteriore parte, quàm poſteriore preſſas & anguſtas. Ac ſimia cynocephalos dentes caninos longiores prominentioreſque, quàm vera ſimia obtinet.* Molarium dentium numerus *in homine incertus eſt, authore* Galeno : *ſæpius enim in utraque maxilla ſunt ſexdecim, interdum viginti, nonnunquam viginti quatuor.* At ſimiæ ſemper certus ac definitus molarium numerus. *Differunt quoque* Maxillares ſimiæ *ab humanis, in figura externa, & radicum numero, quamvis enim priores duo molares ſimiæ, ab humanis, aut nihil, aut certè parum diſcrepent, quia in ſimia primus inferior unum tantùm apicem obtinet : Attamen in poſteriorum dentium menſis, ſecundum longitudinem maxillarum, profunda admodum linea exſculpta eſt. Quam lineam altera etiam tranſverſa, quæ in quinto dente ſimiæ non caudatæ gemina eſt, intrinſecus & extrinſecus ad gingivam uſque protractata interſecat. Quo fit, ut ſinguli ejuſmodi dentes eminentius, ut plurimum quatuor in angulis (nam quinto ſex ſunt) tres verò foveas in medio habentes, duarum ſerrarum mutuò ſibi occurrentium modo committantur : quod profectò accuratiſſius Author* Galenus *explicare minimè prætermiſiſſet, ſi molares ſimiarum deſcripſiſſet.*

Os Hyoides (r) *humano firmè ſimillimum exiſtit, præterquam quod medium ipſius oſſiculum amplius eſt, quàm in homine, & poſteriore ſede inſigniorem oſtendit cavitatem, gibbis ipſius laryngis partibus invehitur, fitque propugnaculum cartilaginis ſcutiformis. In illa enim lata oblongaque apophyſi, interiores partes Oſſis hyoïdis efformat, quæ deorſum adeo producitur, ut inſtar clypei cartilagini thyroïdi obtendatur.*

(a) The

(*a*) The *Cranium* of our *Pygmie* was round and globous, and it feemed to be three times as big as the Head of a Common *Monkey*; for, that I might the better compare them, I procured the *Sceleton* of a *Monkey*, which I found was exactly the length of that of our *Pygmie*: though at the fame time we fhall fee, that in feveral of the Parts, 'twas vaftly different. For meafuring the Head of our *Pygmie* by a Line drawn round from the Nofe, over the *Orbit* of the Eyes, to the *Occiput* or hinder part of the Head, and fo to the Nofe again, I obferved 'twas Thirteen Inches. The *Cranium* of the *Monkey* meafured there only Nine Inches and a quarter. The girth of the Head of the *Pygmie*, from the *Vertix* round by the Ears to the *Vertix* again, was Eleven Inches and an half: in the *Monkey* 'twas only Seven Inches and an half. The longitudinal *Diameter* of the *Cranium* of the *Pygmie* was Four Inches; of the *Monkey* Two Inches and a quarter. The latitudinal *Diameter* of the *Cranium* of the *Pygmie* was Three Inches and above a quarter; of the *Monkey* a little above Two Inches. The profundity of the *Cranium* of the *Pygmie*, from the *Vertix* to the *Foramen* where the *Medulla Spinalis* paffes out, was about Three Inches and a quarter; in the *Monkey* Two Inches. So that in the largenefs of the *Cranium*, the *Pygmie* much exceeds the *Monkey*, as alfo *Apes*, and more refembles a *Man*.

(*b*) The *Sutures* in our *Pygmie* perfectly refembled thofe in an *Humane Cranium*; The *Sutura Coronalis*, *Sagittalis*, and *Lambdoides* being all *ferrated* or *indented* very curioufly, as in *Man*. In the *Lambdoidal Suture* I obferved Nine *Offa triquetra Wormiana*. In the *Cranium* of a *Monkey* I found the *Coronary Suture* for the moft part to be *Harmonia*, and only for a little fpace to be *ferrated* towards the middle, where it meets the *Sagittal Suture*. The *Sagittal Suture* here was indented throughout. The *Lambdoidal Suture*, as extended from the *Sagittal* of each fide for about half an Inch, was *ferrated*; then the *Suture* difappeared, and there was formed here a rifing ridge of the *Cranium*, which was continued to that *Apophyfis* which makes the hinder part of the *Os Zygomaticum*. There was no fuch bony ridge in the *Cranium* of the *Pygmie*. In the *Monkey* too I faw the *Squammous Sutures* very plain, tho' *Riolan* denies them; which likewife in our *Pygmie* was very apparent. Our *Pygmie* therefore in the Structure of the *Sutures* exactly refembled a *Humane Cranium*, and more than *Apes* and *Monkeys* do: For in them the *Coronary* and *Lambdoidal Sutures* were only in part *ferrated*; and they had no *Offa triquetra Wormiana*.

(*c*) In our *Pygmie* there was an *Os Cribriforme*, as in *Man*; 'twas about half an Inch long, and a quarter of an Inch broad; in it I numbred about Thirty *Foramina*; here was likewife that long ridge, (which is call'd *Crifta Galli*) as in a *Man*, to which the *Dura Mater* was faftened.

(*d*) In

(*d*) In the *Cranium* of the *Monkey* there was no *Criſta Galli* ; and where the *Os Cribriforme* ſhould have been, there was a hollow Paſſage which led towards the beginning of the Noſtrils, at the end of which there might be a ſmall *Os Cribriforme* perforated with four or five holes. But this Part appeared very different from the Structure of a *Humane Skull*, as likewiſe from our *Pygmie* ; which was occaſioned chiefly by the great bunching in of the Bones of the Orbit of the Eye, tho' our *Pygmie* too had theſe Bones more protruded in, than they are in a *Humane Cranium*.

(*e*) This appeared more in the *Cranium* of a *Monkey* than in our *Pygmie* ; tho' here too 'twas ſomewhat more than in a *Man's Skull*.

(*f*) The *Sella Equina* in our *Pygmie* was exactly like a *Man's*. In a *Monkey* I obſerved it more riſing and higher. In the middle of the *Sella Turcica ſeu Equina* of our *Pygmie*, I obſerved a *Foramen* ; and the ſame I found in a *Humane Cranium* I have by me.

(*g*) In our *Pygmie* I did not obſerve thoſe two *Cavities* under the *Sella Turcica* which are to be met with in a *Humane Skull*. But the Bone here was very ſpungy and cavernous, and might anſwer the ſame end, tho' not formed perfectly alike.

(*h*) Theſe Cavities in our *Pygmie*, were nothing ſo large as they are in a *Monkey*, but conformable to the Structure of this Part in an *Humane Skull*. And in our *Pygmie* too, I obſerved the *Pterigoidal Proceſſes* as they are in *Man*, but I did not find them in the *Monkey*.

(*i*) The *Oſſa Braguatis* and *Temporum* in our *Pygmie* were very plainly diſtinguiſhed by an *indented Suture*. In the *Cranium* of the *Monkey* theſe Bones were divided by a *lineal Suture* call'd *Harmonia*.

(*k*) The *Maſtoid* and *Styloforme Proceſs* in our *Pygmie* were very little, yet more than in the *Monkey* ; but herein our *Pygmie* rather imitates the *Ape-kind*.

(*l*) Becauſe I would not ſpoil the *Scelcton*, I did not examin the Organ of the *Inward Ear :* But am wholly inclined to *Riolan*, who tells us he always found thoſe Three little Bones, the *Malleolus*, *Incus*, and *Stapes* there ; and no doubt but they are to be met with in our *Pygmie*. Tho' *Caſſerius* therefore thinks *Galen* does not mention them, and never obſerved them, becauſe they are not to be found in *Apes :* But *Riolan* tells us the contrary.

(*m*) The *Os Zygomaticum* in our *Pygmie* was not half ſo big or large as in the *Cranium* of the *Monkey* ; herein therefore our *Pygmie* more reſembles a *Man*.

(*n*) Tho'

(*n*) Tho' the Face of our *Pygmie* was rounder than an *Ape's*, as that is than a *Monkey*, and a *Monkey's* more than the *Cynocephalus*, yet 'twas not altogether fo much as a *Man's* ; the upper Jaw being proportionably longer and fomewhat more protuberant. The Bones of the *Nofe* too in our *Pygmie* more refembled the *Ape-kind*, than the *Humane*, being flat and *finous* ; hence *fimia* ; and not protuberant and rifing as in Man.

(*o*) The *Suture* of the *Palate* in our *Pygmie* was juft the fame as in a *Man*. In a *Monkey* I obferved that peculiar *Suture Riolan* mentions, but did not find it in the *Pygmie* : Only in the *Palate* of the *Pygmie* I obferved a *Suture*, not from the *Dens Caninus*, as was in the *Monkey*, but from the Second of the *Dentes Incifores*.

(*p*) In our *Pygmie* the under Jaw was perfectly clofed at the *Mentum*, as 'twas in the *Monkey* ; and 'tis fo in a *Man*. *Galen* (83) tells us, Ἀπάντων γὸ τῶν ζώων ἄνθρωπ۞ ἔχει βραχυΐατίω τίω γϊυω, ὡς πρὸς τίω ἀναλογίαν διλῶντι τῦ παντὸς σώμαῖ۞· εἶθ᾽ ἑξῆς ἀνθρώπω πίθηκος, εἶτα λύγκας, ᴋᴕ σάτυρςι, κάπιιϑ᾽ ἑξῆς κυνοκέφαλοι. i. e. *That of all Animals a Man hath the fhorteft Chin, or under Jaw, in proportion to his Body* ; *then next to a Man, an Ape, then a Lynx, then* Satyrs, *and after thefe the* Cynocephali. And I may add, of all *Apes*, our *Pygmie* hath the fhorteft. The Articulation of the under Jaw in our *Pygmie* was *Condyloides*, as 'tis in *Man* ; and not *Gynglymoides*, as *Volcherus Coiter* and *Barthol. Euftachius* obferve.

(*q*) Our *Pygmie* had in each Jaw before, four *Dentes Incifores* ; then following them, of each fide a *Dens Caninus* ; then after them of each fide, Four *Dentes Molares*, in all Fourteen Teeth in each Jaw, in both Twenty eight. But our Subject being young, I obferved that all the *Teeth* were not perfectly grown out of the Jaw-bone, and could perceive fome of the *Molares*, that ftill lay hid there, or were not much exerted. In a *Monkey* in each Jaw there were two *Dentes Incifores* before ; then four *Dentes Canini*, two of each fide ; then eight *Dentes Molares*, four of each fide. The Number of the Teeth in each Jaw, and in the whole the fame as in the *Pygmie* : only the *Monkey* had four *Dentes Canini* in each Jaw, the *Pygmie* had but two, as in a *Man* : Or at leaft in the *Monkey*, the two firft of the *Canini* feemed to be Amphibious, between an *Incifor* and *Caninus* ; being not fo broad as the two firft *Incifores*, nor fo much exerted or extended as the two other *Canini* were. In the number of the *Teeth* our *Pygmie* imitated more the *Ape-kind* than the *Humane* : But in the Structure of them , more the *Humane* than the *Ape-kind* ; for the *Menfa* or *Superficies* of the *Molares*, was not fo ferrated as the *Monkey's*, but liker *Humane Teeth*.

I have omitted the Printing the next *Paragraph* in *Riolan*, becaufe I

---

(83) *Galen de Anat. Adminiftr.* lib. 4. cap. 3. p. 94.

K   would

would not be tedious : And for the same reaſon, do not here particularly deſcribe each Bone in the Head and Jaws of our *Pygmie* ; for where
I do not remark otherwiſe, 'tis to be underſtood, that all thoſe Parts are
the ſame in a *Man*, our *Pygmie* and the *Ape-kind*.

(r) There was nothing particular that I obſerved in the *Os Hyoides* of
our *Pygmie* that was different from that of a *Man*'s.

C A P. III.

De Spina & Oſſibus & Adnexis.

SIMIÆ (a) Cervix *brevis eſt, ſeptem vertebris extructa , corpora vertebrarum anteriori parte non ſunt rotunda ut homini, ſed plana. Poſticæ
apophyſes ſpinoſæ non ſunt longæ, & bifidæ, ſed breves, ſimplices, & acutæ.*
In prima (b) *vertebra, nullum ſpinæ veſtigium apparet, imò nulla ſentitur
aſperitas, in anteriore parte corporis primæ vertebræ humanæ obtuſa quædam
eminentia apparet, quæ in ſimia magis extuberat, & in mucronem producitur.
Quod ſi vertebras & ſpinas breves habet ſimia,* (c) *apophyſes tranſverſas
obtinuit longiores, atque ad anteriora magis, quàm in homine reflexas. Imprimis verò ſexta colli vertebra, quæ hunc proceſſum præ cæteris inſignem ad-
epta eſt, eumque bifidum, magiſque recurvum & aduncum, quàm in aliis
vertebris. Hic autem ſpondylus ſextus maximus eſt, propter illas tranſverſas
apophyſes grandiores, in ſimia caudata minor eſt. Septimi ſpondyli tranſverſæ apophyſes ſimplices & tenues, in caudata ſimia bifidæ, & ſatis longæ
exiſtunt, quæ licet in homine ſimplices appareant , ſexto tamen craſſitie non
cedunt.*

(d) *Prima ſimiæ vertebra ad finem proceſſus tranſverſi aſcendentis utrimque foramen habet, ad nervum tranſmittendum, quo humana caret vertebra,
ſeptima colli vertebra in homine ſæpius eſt perforata : Unde evenit, quod
tranſverſi proceſſus hujus vertebræ non ſunt ſimiles apophyſibus tranverſis colli, ſed potiùs thoracis apophyſibus tranſverſis aſſimilantur.* (e) *Vertebrarum dorſi corpora parum ab humanis differunt, neque apophyſes multùm diſ-
ſimiles ſunt, exceptis rectis ultimarum duarum vertebrarum,quæ rectiores ſunt
in ſimiis,paulum deorſum inclinat in hominibus.* In (f) *poſtremis vertebris
dorſi reperiuntur quatuor infernæ apophyſes articuli gratiâ conſtructæ. In
humanis vertebris duæ tantùm notantur, quas etiam in lumborum vertebris
obſervabis.* (g) *In ſimia decima dorſi vertebra, infra ſuprave ſuſcipitur, at
in homine eſt duodecima.*

(h) *Lumbi,*

(h) *Lumbi, inquit* Galenus, *in ſimiis ſunt longiores quàm in hominibus, ſi pro ratione reliquarum partium hoc æſtimare velis, nam in homine quinque vertebræ lumbos efformant, in ſimiis non ſecus, quàm in aliis quadrupedibus ſex adſunt.* (i) *Harum vertebrarum proceſſus ab humanis differunt. Siquidem tranſverſi in homine teretes ſunt & oblongi, nonnihil in exteriora converſi, coſtularum vicem gerentes. In ſimia ſunt ampli, intrò ſpectantes, & inſtar ſquammæ tenues: figurâ caudam hirundinis referunt, aut cornu retortum, quod oblongo acutoque mucrone erigitur, ac ſurſum vergit. Ac tertia lumbi vertebra primò incepit tranſverſum conſequi proceſſum, qui brevis eſt. Reliqui ſubſequentes longiores exiſtunt.* (k) *Poſteriores proceſſus ſpinoſi non ſunt recti, ſed ſupernè ſpectant, atque excipiuntur à ſupernis incumbentibus vertebris, quæ hiatu ſive ſciſſura triangulari inter duos tranſverſales proceſſus exiſtente, dum in poſteriora ſimia ſpinam inflectit, eoſdem tranſverſales excipiunt.*

(1) *Obſervandum venit in homine circa radices infernas tranſverſarum apophyſeon lumbarium, atque etiam duarum infirmarum thoracis, quædam tubercula magnitudine figuraque meſpilorum nucleos referentia ſæpius reperiri, quæ cum in canibus & ſimiis non habeantur, ſuſpicari poſſet aliquis vicem illarum quas paulò antè deſcripſi proceritatum in homine tenere.*

(m) Os Sacrum *ſpinæ fundamentum in homine, ex tribus vertebris conſtatur: In ſimiis ex duabus tantum componitur, quibus ilium Oſſa copulantur.*

(n) *Simiæ longior eſt* Coccyx, *quàm homini, pluribus ideo conſtructus Oſſibus, quæ juxta commiſſuram Oſſis Sacri perforata ſunt, medullamque continent, atque nervos antrorſum & retrorſum emittunt, quæ omnia deſunt in coccyge hominis: cur autem ſimia vero coccyge caruerit, rationem reddit* Fallopius, in Obſervat. Anatomicis.

(o) Homo, *inquit* Galenus, *ex omnibus animalibus* Coſtas *curviſſimas habet, propterea latiſſimum pectus obtinuit. Simiæ latius cæteris pectus datum, ſed humano anguſtius.*

(p) *Porro ſimia, tam caudata, quàm non caudata, coſtas viginti ſex præ ſe fert, cum in homine tantum viginti quatuor reperiantur. Harum utrimque ſunt octo veræ coſtæ, in homine ſeptem, quæ per articulum ſterno committuntur. Quinque vero nothæ coſtæ non deſinunt in perfectam cartilaginem verſus ſternum invicem conſtrictæ, ut in homine, ſed oſſeæ magis quàm cartilaginoſæ, à ſe mutuò disjunguuntur. Coſtæ in ſimia, tam caudata, quàm non caudata, ſpatiis vertebrarum intermediis inſeruntur: at in hominibus magis cerporibus vertebrarum attexuntur.*

(q) Sternum *octo conſtat Oſſibus rotundis, quorum primum aliquantiſper prominet, ſupra cartilaginum duarum primarum conjunctionem, quæ duæ cartilagines videntur amplexari ſuperiore parte primum os ſterni. Cartilagines*

coſtarum

coſtarum commiſſuris Oſſium ſterni accreſcunt, duæ ultimæ concurrunt ſimul in articulationem ultimi & penultimi oſſis ſterni. Ultimum os ſterni xiphoidem cartilaginem referens, impensè longum eſt, & teres.

(r) *Simia quoad ſcapulas & claviculas homini maximè ſimilis eſt*, authore Galeno. *Clavicula incipiens à primo ſterni oſſe ad medium coſtæ rectè procedit, inde ad acromion uſque multùm curvata intumeſcit : huic articulationi interjectum eſt oſſiculum, quod in homine ad decimum octavum annum & ultrà, appendix exiſtit : at in ſimia, nec iſtud oſſiculum, nec illius veſtigium nullum apparet, imò pars illa robuſtiſſima eſt.*

(a) In our *Pygmie* there were ſeven *Vertebræ* of the Neck, as there are in a *Man*, and an *Ape* too ; but they were ſhort, making in length about two Inches ; and ſeemed more to imitate thoſe in *Apes*, being flatter before, and not ſo round as in *Man*. And their *Spines*, tho' they were longer, and more obtuſe, and not ſo acute as in *Monkeys* ; yet they were not *bifide*, as they are in *Man*.

(b) In the firſt *Vertebra* of the Neck in the *Pygmie* there was no *Spine*, but an Aſperity ; in a *Man* there is a *ſmall Spine*. And before, 'twas like to the *Humane*, having an Obtuſe Eminence, and not running to a *Mucro*, as in the *Ape* and *Monkey*. The *Dens* of the ſecond *Vertebra* in the *Pygmie* was partly *Cartilaginous.*

(c) I did not obſerve in the *Pygmie* the *Tranſverſe Apophyſes* to be longer, nor to be reflected more forward, nor the *ſixth Vertebra* to be larger than the others ; nor the *ſeventh Vertebra*, to be any thing different from the ſame in *Man* ; but in all theſe Circumſtances, our *Pygmie* ſeemed to imitate the Structure of the ſame Parts in *Man*, more than does the *Ape-kind.*

(d) Thoſe *Foramina* obſerved in the *Vertebræ* of the Neck of *Apes*, were wanting in our *Pygmie*, who herein imitated the *Humane Sceleton.*

(e) I did not obſerve any difference between the *Vertebræ* of the *Back* of our *Pygmie* and thoſe of a *Man's* ; nor what *Riolan* remarks of the *Apophyſes rectæ* of the two laſt *Vertebræ.*

(f) In the lower *Vertebræ* of the Back of the *Pygmie*, I obſerved but two *Apophyſes infernæ*, as 'tis in a *Humane Sceleton :* in a *Monkey* there are four *Apophyſes* there.

(g) Our *Pygmie* if *Riolan's* account be true, is different both from the *Ape* and *Man* too ; for here 'twas the thirteenth *Vertebra, quæ infra, ſuprave ſuſcipitur.*

(h) The

(*h*) The *Vertebræ* of the *Loins* in our *Pygmie* were about two Inches long; and their number the fame, as in a *Man, viz.* five; and not fix, as are in *Apes* and *Monkeys :* But the *Os Ilium* of each fide does afcend fo high, as to include the two lower *Vertebræ*; which is not fo in *Man.*

(*i*) The *Tranfverfe Proceffes* of the *Lumbal Vertebræ* in the *Pygmie,* were round and thick, as in *Man*; and not thin and flat, or broad, as in the *Monkey.*

(*k*) The *Spines* of the *Lumbal Vertebræ* in the *Pygmie* , were ftrait, as in a *Man*; and not bending upwards, as in the *Ape* and *Monkey-kind.*

(*l*) I am apt to think thefe *Tubercula* are in our *Pygmie*; but our Subject being young, and feveral of the Parts not yet hardened into *Bones,* but *Cartilaginous*; I was not fully fatisfied herein, and do leave it as a *Quære.*

(*m*) *Riolan* in this account is miftaken, nor is he here confiftent with himfelf, as to what he writes of this part in other places. *Joh. Philip. Ingraffias* (84) who has wrote a moft learned and incomparable *Comment* upon *Galen's* Book *de Offibus,* tells us; *Amplum Sacrumve Os in Homine fex vel ad minus quinque ex Offibus confiat.* Galenus *tamen, Simiarum Caunmve Sceletos diffecans, interdum quatuor ex Offibus componi inquit. Sub Offe inquam Sacro largius fumpto, Coccygem quoque comprehendens :* (quem *Coccygem pro uno Offe , ut in præfenti textu facit, tanquam quartum adjungens affumpfit, à Sacro interim diftinguens-) fæpius autem tribus duntaxat proprie fumptum Os Sacrum à Coccyge diftinctum expreffit, uti nunc etiam facit : unde tribus ex partibus conftructum effe ait, tanquam ex propriis Vertebris.* So *Falloppius* and others do make the *Os Sacrum* in a *Man* to confift of fix Bones, fometimes five. In our *Pygmie* the *Os Sacrum* was compofed of five Bones : But in the *Sceleton* of a *Monkey* I obferved but three Bones or *Vertebræ* which did make up the *Os Sacrum.*

But as our *Pygmie* in the number of the *Vertebræ* which compofes the *Os Sacrum,* did imitate the *Humane kind*; fo in other refpects 'twas much liker to the *Sceleton* of *Apes* and *Monkeys :* For the *Os Sacrum* here, was nothing fo dilated and fpread, as 'tis in *Man*; but contracted and narrow as 'tis in *Apes* ; and very remarkably different from the *Humane Sceleton*; as 'twas likewife in the *Spines* and *Proceffes* which more refembled the *Ape-kind.*

(*n*) The *Os Coccygis* in our *Pygmie* confifted of four Bones, as 'tis in an *Humane Sceleton,* and thefe not perforated. In the *Ape,* and efpecially in the

---

(84) *Comment. in Galen. de Offibus,* Cap. x. Text. 3. pag. m. 184. -

No...

Monkey, there are more Bones, and thoſe perforated, as *Riolan* deſcribes them.   Hence *Julius Pollux* ſtiles it , ὁ τρητὸς κόκκυξ, *Perforatus Coccyx*. This *Os Coccygis* makes a little bunching out of the Skin in the *Pygmie*, as I have repreſented it in my *ſecond figure*, and is remark'd before (*vide* *pag.*14.)but in *Man*,'tis not protuberant.   What *Riolan* obſerves out of the *Nubian Geography*, of a Nation in the Iſle of *Namaneg*, having Tails, I think is fabulous; unleſs they be *Monkeys*, or of that kind : I am certain that Story of the *Kentiſh Longtails* he mentions, is utterly falſe, tho' he modeſtly expreſſes himſelf, *fabuloſum puto*.   His words are theſe : *In Inſula Namaneg Maris Orientalis, Gens eſt caudata*, ex Geographiâ Arabicâ Nubienſi, pag. 70. *fabuloſum puto quod de Anglis Caudatis referunt Hiſtorici, quibus ob injuriam* D. Thomæ Cantuarienſi *illatam, Deus Coccygem inſtar Caudæ produxit* (85).

(o) In our *Pygmie* the *Ribs* were altogether as much curved, as in an *Humane Sceleton* ; and it was as full *cheſted* as a *Man*.

(p) In the number of the *Ribs* our *Pygmie* imitated the *Ape-kind :* for it had thirteen of a ſide, ſix and twenty in all : In a *Man* there is but twenty four, tho' ſometimes there has been obſerved thirteen of a ſide. As to the other Particulars that *Riolan* mentions, *viz.* the number of the *Coſtæ veræ*, and the oſſious Extreams of the *Nothæ*, and the Articulation of the *Ribs*, herein our *Pygmie* more reſembled a *Man :* for it had but ſeven *Coſtæ veræ* that were articulated to the *Sternum* ; and the Extreams of the *Nothæ* were *Cartilaginous*, not *Oſſious*, and continued to the *Sternum* as in an *Humane Sceleton* ; and the Articulation of the *Ribs* was more on the Body of the *Vertebræ*, than in the *Interſtices*.   *Drelincourt* is miſtaken in mentioning but twelve *Ribs* in the *Ape*, of a ſide, or his was different.

(q) *Jo. Philippus Ingraſſias* (86) makes eight Bones in the *Sternum* of Infants ; and tells us, that in time theſe Bones do coaleſce, and grow fewer.   *Galen* makes ſeven Bones in the *Sternum* , according to the number of the *Coſtæ veræ* that inſert their *Cartilages* into them.   But the eighth Bone *Ingraſſias* ſaith, is for the *Cartilago Enſiformis*.   In the *Sternum* of our *Pygmie* I numbred ſeven Bones, the two laſt being ſmall and partly *Cartilaginous* ; and here the *Cartilages* were inſerted at the Commiſſures and Joynings of the Bones of the *Sternum*.   The *Cartilago Enſiformis* was long and roundiſh.   The whole of the *Sternum* of our *Pygmie* much more reſembled the *Humane Sceleton*, than the *Monkey's*, being much broader and larger, and as far as I obſerved juſt alike.

(r) The *Scapula* of our *Pygmie*, tho' in moſt reſpeɛts it reſembled a *Man*'s, yet I thought it did not ſo much, as a *Monkey's* ; for it ſeemed

(85) *Riolan. Encheirid. Anat.* lib. 6.cap. 16. p. 451.   (86) *Jo. Phil. Ingraſſias Comment. in Galen. de Oſſibus*, Cap. 12. Text. 1. pag. m. 190,

narrower,

narrower, and the *Basis* was proportionably longer. But this I suppose might happen in preparing the *Sceleton* by paring away the *Cartilages* (for the Creature was young) which in a longer time would have hardened into a Bone. So likewise that *Procefs* which receives the *Clavicula* call'd *Acromion*, was *Cartilaginous*, as was likewise the End of the *Procefsus Coracoides* , and of the *Cervix* it self, which laft received the head of the *Shoulder Bone*. So that as yet there was not a *Sinus* formed here for the receiving it; but that Extream was flatter than ufually and plain ; nor was there that *Sinus* under the *Spine*, as in an adult *Humane Scapula*.

I obferved no difference in the Figure and Structure of the *Clavicula* in our *Pygmie* and in a *Man*. Nor did I obferve that Bone *Riolan* mentions, but a large *Cartilage* which did conjoyn that Extream of the *Clavicula* to the *Acromium*, which in time might become long ; this *Cartilage* was about a quarter of an Inch long.

C A P. IV.

De Artubus Superioribus.

(a) SIMIÆ *& Hominis Omoplatæ omnino fimiles funt*.

(b) Humerus *fimiæ non admodum ab humano differt, in caudatæ diffimilis eft juxta inferius caput, quod cubito articulatur. Hac enim regione reflectitur ab exteriori parte introrfum, atque in illa flexura canaliculum acquirit ex oppofito latere pervium.*

(c) Cubiti Offa duo *in utraque fimia humanis refpondent.*

(d) Carpus *fimiæ non valde differt ab humano, obtinuit tamen nonum os, de quo fic loquitur* Euftachius : *Hoc ofsiculum non in prima brachialis acie eft locatum, fed tertio ejufdem acici ofsi incumbit, atque inter ea quæ indicem & medium digitum fuftinent feipfum inferit: vocatur à* Galeno διχάς, *hoc offe videntur carere fimiæ caudatæ, fed ejus loco adipifcuntur os peculiare, quod carpi ofsi cubito fubftrato annectitur, & fatis longè protuberat. Deinde inftar cornicis verfus manus volam incurvatur , atque cum procefsu ofsis carpi radio articulati, magnam cavitatem mufculorum tendinibus efformat.*

(e) Metacarpij, Digitorumque Offa *fimiæ, tam caudatæ quàm non caudatæ, parum admodum ab humanis ofsibus difcrepant. Simia quidem magnum manus digitum* Pollicem, *mutilum obtinet, & curtum, & indici propinquum, non oppofitum, inftar alterius manus, ut in homine : Reliqui digiti multò funt minores digitis pedum.*

(*a*) I

(*a*) I have already mentioned that the *Scapula* or *Omoplata* in our *Pygmie* did not ſeem ſo like a *Man*'s, as a *Monkey*'s did ; nor does it appear ſo in my *figure* ; not but that I think 'tis ſo, when adult ; and it's *Cartilages* are hardened into a *Bone*: but my *figure* only repreſents what was now formed into a *Bone*, and without the *Cartilages*, which in time would have become bony.

(*b*) The *Os Humeri* in our *Pygmie* was a little above five Inches long, juſt the ſame length with the *Thigh Bone*, and not altogether ſo thick. That end which was joined to the Bones of the *Cubit*, was about an Inch and a half broad. I obſerved here, upon the flexure of the *Cubit* forwards, that in the *Os Humeri* there was a deep *ſinus*, and the Bone ſo thin here, that it would admit the Rays of Light thorough ; but 'twas not pervious as *Riolan* ſaith it is in a *Monkey* ; nor did I obſerve it ſo, in the *Sceleton* of a *Monkey*.

(*c*) In the *Pygmie* the Bones of the *Cubit* were exactly like a *Man*'s. The *Ulna* was five Inches long ; the *Radius* five Inches and an half. They had large *Cartilages* at both Extreams.

(*d*) So likewiſe the Bones of the *Carpus* in the *Pygmie* reſembled thoſe of a *Man*. I did not obſerve here that *ninth Bone* deſcribed by *Euſtachius*. For indeed in our Subject, there were but four in each *Carpus*, that were *oſſified* : the others were only *Cartilaginous*.

(*e*) In the *Hand*, our *Pygmie* reſembled the *Ape* and *Monkey*-kind. For tho' the Bones of the *Metacarp* and *Fingers* were like to thoſe of a *Man*; yet the *Thumb* was much ſmaller, than the other Fingers, and ſhorter, and liker the *Ape-kind*. This *Galen* frequently takes notice of. 'Tis true, the other *Fingers* were much larger in our *Pygmie* than in the *Ape-kind*, and more reſembling thoſe of a *Man*, ſo that I was ſurpriſed to ſee them ſo big : but the *Thumb*, which the *Ancients* and *Galen* call ἀντήχειρα, and *Hippocrates* μέγαν, in our Subject was ſo diſproportionate and little, that as *Galen* remarks (87), any one that ſhould view it, would think that it was but a *ridiculous imitation* of *Man-kind*, and nothing anſwering to it's Names. And in the precedent Chapter he vigorouſly diſputes againſt the *Epicureans* and the *Followers* of *Aſclepiades* ; and from the admirable Structure and wiſe Contrivance of all the Parts, and particularly the *Tendons* that go to the *Fingers* ; he confutes their *Hypotheſis* as vain, and hath this noble *Epiphonema*, ὅτι, ὧ πρὸς Διὸν εδὶν ἐχενίες ἐν τοσαύταις χαλαφύσαν μέμψασθαι, οὖτ ἐν τ ὅπον τὴ πνόνίων, ἐπι τ τόπον, οὖτε τ τρίπον τῆς ἐμφύσεως, ἀλλ' ἐν ἀπάσαις οὗπαις Θαυμασιὼ ἀναλογίαν ἐρῶνίες, μιᾶς μώνης ἀπολυμένην, κατὰ τ αὐτὸν τρίπον ἐν ἀμφοῖς ταῖς μεγάλοις δακλύλοις, ᾗ ταύτης ἐκ ἀλδ.

(87) *Galen de uſu Partium, lib. 1. cap. 22. p. m. 310.*

χὼς,

γως, ἀλλ' ὅτι μηδὲν αὐτῆς ἐχρησόμεν, εἰκῆ φάτι, ᾗ χωρὶς τέχνης, ἅπανΊα τὰ. τοιαῦτα γεγονέναι. i. e. *Vos, per Deos immortales, cùm nihil habeatis, quod in tot Insertionibus reprehendatis, neque Tendonum molem, neque locum, neque Insertionis modum, sed in his omnibus mirabilem quandam Proportionem videatis, unâ solâ in utroque magno digito similitèr perditâ (& hoc non sine ratione, quod eâ non egebamus) temerè dicitis & absque Arte omnia hujusmodi facta fuisse.*

The Bones of the *Metacarpus* in the *Pygmie* were an Inch and three quarters long. The two laſt Joints of the *Thumb* were ſcarce an Inch long ; the firſt Joint of the *Thumb* was a little above an Inch. The *Fore-finger* was two Inches and almoſt an half : The *middle Finger*, two Inches and three quarters. The third or *Ring-finger* was two Inches and half a quarter ; and the *little Finger* was not full two Inches long. The firſt *Joint* of the *fore* and *middle Finger* was above a quarter of an Inch broad, and the Girth of each about was an Inch. The *Pygmie* therefore in the *Fingers,*having them ſo large and thick, imitated a *Man* ; but in the *Thumb*, which was ſo ſlender and ſmall, it reſembled the *Ape-kind.*

C A P. V.

De Artubus Inferioribus.

O S S A (a) Ilium *in utraque simia, tam caudata quàm non caudata,toto habitu, & figura distant ab humanis : dehiscunt enim eo in loco, ubi pubis Ossa esse debebant, atque omnino privantur Osse pubis : propterea ad velociter currendum ineptæ sunt. Ischij articulus planè dissimilis est ab illo hominis, ut notavit* Galenus.

(b) *Ea est* Femoris ſtructura *in simia, ut eam stare rectam non permittat, nec instar hominis corpus suum erigere, aut incedere, ne quidem sedere, quia femoris caput obliquius in articulo coxæ, committitur.* (c) *In homine cervix rotundi capitis femoris oblonga est, & sensin obliquè deorsùm ducitur. In simia verò brevis, & propemodum transversa visitur. Sed femoris cervice, apophyses duæ, trochanteres dictæ, in simia similes sunt humanis , verum in illa, ut in caudata minores.*

(d) Patella *utriusque simiæ manifestum discrimen ab humana demonstrat : est enim oblonga, non rotunda. Quamvis autem extrinsecus gibba sit, atque intus cava, nihilominus longè aliter se habet quàm in homine. Nam secun-*

L

dum

*dum ipſius longitudinem recurvatur, excavaturque adeo, ut nihil propemo-
dum in medio emineat, curvo aduncoque ejus ſinu naviculam quandam ele-
gantiſſimè referre videatur. In caudata ſimia patella videtur ex duobus Oſſibus mutuò adnatis conſtructa.*

(e) Tibiæ utrumque Os *in utraque ſimia humanis Oſſibus ſimillimum* eſt.

(f) *Simiæ* Pes *ab humano maximè diſcrepat,* eſt *enim oblongus latuſque
homini, anguſtus breviſque ſimiæ, pro ratione corporis, pediſque digiti longiores ſunt, ſed metatarſi Oſſa breviora, calcaneum verò anguſtius, & anteriori
in parte qua cum Oſſe cyboide committitur, latius evadens, magiſque inibi
longum, quàm retrò, impedit ne ſimia diu erecta, & ſtare, & ambulare queat.
Aſtragalus* Galeno *tenuis non efficitur, ſed manens ſublimis , Oſſi ſcaphoidi
conjungitur, quod ſimiæ repugnat, in qua aſtragalus humilem, oblongam atque
tenuem cervicem habet. Planta in ſimia ex quatuor Oſſibus componitur. Pollex ex tribus, inquit* Euſtachius. *Quamvis* Volcherus *in caudata ſimia
nullam obſervarit differentiam, quà diſcreparent ab homine.* (g) *Digitorum
notiſſima* eſt *diſcrepantia in homine, ut notavit* Galenus, *omnes una ſerie
diſponuntur, breviſſimoque ſpatio diſcreti , multo minores ſunt, quàm qui in
manu habentur. Nam quantò pes ſumma manu major* eſt *, tanto illius digiti
manus digitis ſunt minores.*

(h) *Accedit quod pollex longitudine indici æqualis* eſt, *quem dupla craſſi-
tudine ſuperat, taliſque* eſt *quatuor digitorum commenſuratio, ut ab indice ad
minimum ſemper deficiat longitudo : & ſecundæ aciei Oſſa, ſi indicem exceperis, breviora ſunt iis, quæ in tertia phalange reponuntur. Hæc omnia in
utraque ſimia aliter ſe habent, omnes enim pedis digiti inſigni ſpatio diſcreti
ſunt, multoque longiores, quàm in manu exiſtunt : Pollex cæteris digitis bre-
vior tenuiorque* eſt, *atque diverſam ab aliis poſitionem ſortitur, dehiſcit etiam,
ut pollex in manu valde ab indice. Digiti pedis ſimiæ, manus humanæ digitorum ſeriem imitantur,* eſt *enim pollex in pede ſimiæ reliquis digitis bre-
vior, inter alios quatuor digitos ut in manu, medius omnium longiſſimus.*

(a) There was no Part I think in the whole *Sceleton* where the *Pygmie* differed more from a *Man,* than in the Structure and Figure of the *Os
Ilium :* for in a *Humane Sceleton* thoſe Bones were ſpread broad, forming a
Sinus or Hollow on the Inſide. In the *Pygmie* they were proportionably longer and narrower, and not ſo *Concave* on the inſide, but in all
reſpects conformable to the Shape of the ſame Bones in the *Ape* and
Monkey-kind. But why *Riolan* ſhould deny the *Os Pubis* to be in *Monkeys,* I ſee no reaſon ; for naturally there is not that *Dehiſcence* or Separation of the *Os Pubis,* as *Coiter* has given in his *Figure* of the *Sceleton* of a *Monkey,* and as he deſcribes it ; from whence I ſuppoſe *Riolan*
borrows this Deſcription : for in the *Sceletons* of two *Monkeys* I obſerved
theſe Bones were joined together, and in the *Pygmie* they are cloſed as
in a *Man.* When the *Cartilage* that joins them is divided, they will part
aſunder ;

afunder ; but otherwife they are firmly knit together. This therefore is no reafon, why they fhould not run faft : and the contrary was obferved of the *Pygmie* that it did fo.

The length of the *Os Ilium*, from it's *Spine* to it's Conjunction with the *Os Ifchium*, was three Inches ; where 'twas broadeft, 'twas an Inch and half ; where narroweft, but three quarters of an Inch. The *Os Ifchium* was an Inch and three quarters long ; the *Os Pubis* was an Inch long.

(*b*) I did not obferve any difference in the Structure of the *Thigh-bone* in our *Pygmie* from that in *Man* ; nor was it's *Articulation* or Infertion of it's Head into the *Acetabulum*, more oblique than in *Man*. So that from this *Articulation*, I faw no reafon why it fhould not walk upright and fit ; our *Pygmie* did both : When I faw it, 'twas juft a little before it's death ; and tho' 'twas weak and feeble, it would ftand, and go upright.

The length of the *Thigh-bone* in the *Pygmie* was five Inches : The girth of it in the middle an Inch and three quarters; where 'twas joined to the Bones of the *Leg*, 'twas an Inch and almoft an half broad.

(*c*) The Neck of the Head of the *Thigh-bone* in our *Pygmie* was not different in it's length, as I did obferve, from that of a *Man's*, but the fame proportionably ; as were likewife the two *Apophyfes*, called *Trochanteres*.

(*d*) The *Patella* in our *Pygmie* was not yet *offified*. As much as I could difcover of it's fhape, it was the fame as in *Man* ; round and not long ; and but one Bone, and not two, as *Riolan* defcribes it in the *Monkey*. In the *Sceletons* of the *Monkeys* I ufed, thefe Bones were loft, fo that I did not obferve them.

(*e*) The two Bones in the *Leg*, the *Tibia* and the *Fibula* were juft the fame in the *Pygmie* as in *Man* ; and their Articulations were alike : The *Tibia* was four Inches long ; the *Fibula* was a little fhorter. The girth of the *Tibia* in the middle was about an Inch ; of the *Fibula*, about half an Inch.

(*f*) What makes the *foot* of the *Pygmie* feem different from a *Man's*, is chiefly the length of the *Toes*, and the Structure of the *great Toe*. In other refpects, it has a great refemblance with it. For the Bones of the *Metatarfus* here, feemed proportionably as long as in *Man*. The *Os Calcis*, *Calcaneum* or *Heel-bone* was not narrow, but broad ; and forewards, where 'twas joined to the *Os Cuboide* or *Cubiforme*, not broader, nor longer, than behind ; where it juts out fo far, as fufficiently fecures it's

L 2   ftanding ſtanding or walking erect. The *Aſtragalus* I did not obſerve different from a Man's. The *Scaphoides* or *Naviculare* here was *Cartilaginous*. If one reckons three Joints in the *great Toe*, then there can be but four Bones in the *Planta Pedis*, or *Metatarſus*; which with *Euſtachius* I am more inclined to, becauſe really this Part performs upon any occaſion the uſe of an *Hand* too ; and the *great Toe*, (like the *Thumb* in the *Hand*) ſtands off from the range of the other Fingers. Beſides, I obſerved a difference in the Colour in the Bones of the *Metatarſus* and the *Toes* : for the Colour of the *Toes* was white and opace ; the Colour of the Bones of the *Metatarſus* was like to that of the *Cartilages*, and more tranſparent. Now all the three Bones in the *great Toe* were of the ſame colour, white as were the other *Toes*. Therefore I ſhall make but four Bones in the *Metatarſus*, anſwerable to thoſe of the *Metacarpus* in the *Hand*, and three Bones in the *great Toe*.

(*g*) And as the *Hand* of our *Pygmie* in ſome Parts reſembled the *Humane* ; in others the *Ape-kind* : So the ſame may be ſaid of the *Foot* too. For the *Heel*, the *Tarſus* and *Metatarſus* were like to the *Humane*. But all the *Toes* were liker to the *Ape* and *Monkey-kind* : For the *Toes* here , if we may call them *Toes*, and not rather *Fingers*, were almoſt as long as the *Fingers* in the *Hand*; much longer proportionally than in Man, and not lying ſo cloſe together: But the *Bones* of the *Fingers* in the *Hand*, were larger and bigger than thoſe of the *Toes*.

(*h*) The *great Toe* in the *Pygmie*, was ſhorter than the firſt of the other Toes ; tho' in a *Man* 'tis altogether as long ; and herein it reſembles the *Ape-kind*. But whereas *Ariſtotle* (as I have remarked) mentions, that in *Apes* the *middle Toe* is the longeſt , as is the *middle Finger* in the *Hand* ; In the *Sceleton* of the *Pygmie* I did obſerve, that the *firſt* and *middle Toe* were both much of a length , each meaſuring an Inch and three quarters : The *third* and *little Toe* were about an Inch and an half long ; the *little Toe* being rather ſomewhat ſhorter than the *third Toe*. If in the *great Toe* you reckon three *Articuli*, as *Euſtachius* does, then from the *Tarſus* to it's Extream, the *great Toe* meaſured two Inches and an half : but if with *Coiter* you make but two *Articuli* or *Joints* in the *great Toe*, and the other to be a Bone of the *Metatarſus* ; theſe two were only an Inch and a quarter long: The four Bones of the *Metatarſus* were much of a length, being about an Inch and a quarter long.

This *great Toe* (as has been already frequently remarked) being ſet off from the range of the others, more reſembles a *Thumb*. This Difference I obſerve in it's make, That the Bones that compoſe it, are much bigger and larger, than any of the other *Toes* ; and in reſpect of the *Thumb* in the *Hand*, vaſtly bigger. In the *Sceleton* of a *Monkey* I did not obſerve the Bones of the *great Toe* , to exceed thoſe of the other. But as the *Thumb* in the *Foot* is much bigger, than that in the *Hand*; ſo the *Fingers* in the *Hand* are much larger than thoſe in the *Foot*.   CAP.

C A P. **VI.**

De Sesamoideis.

IN *Homine* Ossa Sesamoidea *pauca sunt, magnaque ex parte cartilagino-sa, & si ea quæ pollici applicantur exceperis, in constanti sede firmata. In simia verò multa, atque magna occurrunt, & ossea perpetuò sunt. Cuique primo quatuor digitorum internodio, & secundo pollicis gemina ferè semper adnectuntur. Duo officula magnitudine ciceris, supra utrumque tuberculum femoris in origine gemellorum reperiuntur.*

As to the *Ossa Sesamoidea* in our Subject, I have very little to say : For it being young, very likely they might be only *Cartilaginous*, and the Skin adhering so firmly here, they might be taken off with it. Since they are in *Apes*, I do not doubt, but that they were in our *Pygmie* too, tho' I did not observe them.

Having now made my Remarks upon the *Comparison*, that *Riolan*, or rather *Eustachius* and *Coiter*, have given us, between the *Sceleton* of a *Man*, an *Ape*, and a *Monkey* ; and shewn wherein the *Sceleton* of our *Pygmie* either agreed or disagreed from any of them , I shall make some Reflections upon the whole ; and more particularly upon some Parts, which deserve here a more distinct Consideration. But shall first of all take the Dimensions of the *Sceleton*, and of some other Parts I have not mentioned already.

As from the top of the *Cranium* to the Extream of the *Heel* in a strait Line, the *Sceleton* of the *Pygmie* measured about two Foot ; from the first *Vertebra* of the *Neck* to the last of the *Os Coccygis* , eleven Inches ; from the head of the *Shoulder-bone*, to the end of the *middle Finger*, 'twas about fifteen Inches ; the end of this *Finger* reaching in an erect Posture an Inch and half below the *Patella :* whereas in an *Humane Sceleton*, from the end of the *middle Finger* to the lower part of the *Patella*, it wanted five Inches and an half : Our *Pygmie* therefore herein imitated the *Ape-kind*. From the head of the *Thigh-bone*, to the bottom of the *Os Calcis* in the *Pygmie*, was about ten Inches. From the setting on of the first *Rib*, to the fastening on of the last, was four Inches. The distance between the last *Rib*, and the *Spine* of the *Os Ilium*, not full two Inches. From the *Spine* of the *Os Ilium*, to the bottom of the *Os Pubis*, in a strait Line, was four Inches and three quarters. The distance between the end of the *Scapula*, and *Spine* of the *Os Ilium* about three Inches.

Both

Both when it was alive, and after it's death, I admired the ſtraitneſs and ſhape of it's *Back*. Now the *Scapula* coming down ſo low on the Ribs, and inclining towards the *Vertebræ* of the Back, and the *Os Ilium* riſing ſo high, they do contribute very much towards it ; and muſt alſo afford a great ſafeguard and ſtrength to the *Back* and *Spine*.

The *Sceleton* of our *Pygmie* was juſt the ſame length of one of a *Monkeys* that I borrowed : But becauſe I obſerved moſt of the *Apophyſes* of the Bones to be *Cartilaginous* in the *Pygmie*, I muſt conclude, that 'twas but *young*; and that probably it might grow taller ; to what height I am uncertain. Yet I can by no means be induced to believe , that it would ever arrive to the Stature of a *Man*, as ſome ſort of this *Species* of Animals has been obſerved to do ; for then I could not expect, to have ſeen here, the *Bones* themſelves ſo ſolid, or the *Cranium* to be ſo entirely oſſified, or the *Sutures* to be ſo cloſed and indented, and the *Backbone* and *Ribs* ſo fully hardened, as all the Bones of the *Artus* or *Limbs* were likewiſe, except at their *Apophyſes*, and in the *Carpus* and *Tarſus*. Now all theſe Parts that had theſe *Cartilaginous Apophyſes*, had already acquired ſo great a length, in proportion to the reſt of the *Body*, that 'tis not to be imagined, that they would have exceeded it, or at leaſt not much ; and conſidering that *Animals* come to their ἀκμὴ of growth ſooner or later, according to their *Longevity*, as a *Man*, (till he is paſt the Age that any of theſe Creatures, it may be, arrive to) does not leave growing : this inclines me to think, ſince we found moſt parts of the Body ſo perfected here, that it might not in time much exceed the height it had already acquired. I could have wiſhed that thoſe that have wrote of any of this *Species* of *Animals*, had given us their Dimenſions and Ages ; but they are ſilent herein, or at leaſt too general : only *Le Compte* obſerved an *Ape* in the Straits of *Molucca* four foot high ; but this may not be our ſort. As to thoſe of *Borneo*, I was informed by a Sea-Captain who uſed thoſe Parts, that the King there formerly had one as tall as a *Man*, that would frequently come down to the Town, and a great many Stories are told of him. The ſame Captain had two given him, both young, and about the height of our *Pygmie* ; but theſe were not hairy, but naked as a *Man*; and one of them that he carried to *Batavia*, was looked upon as ſo great a Rarity, that all the time he ſtaid there, his Ship was conſtantly viſited by ſuch as came to ſee it. But 'tis Matter of Fact, not Reaſoning, that will beſt determine this doubt , and a faithful Obſervation that muſt inform us, to what tallneſs this ſort of *Animal* in *Angola*, and the Countries thereabout, does uſually grow ; for in different Countries they may be different in this reſpect, tho' the ſame *Species*, as is ſeen even in *Mankind*.

'Tis not therefore that I am fond of the word *Pygmie*, that I have call'd our *Animal* ſo, or that I would undertake to juſtifie our preſent
Subject

Subject to be exactly the *Pygmie* of the Ancients : Of this *Quadru-manus* sort of Animals there are divers *Species*, and some may be taller and others shorter ; but all of them being but *Brutes*, I was unwilling to call ours a *Man*, tho' with an *Epithet*. 'Twas necessary to give it a Name, because not tallying exactly with the Descriptions of those that are given us, I did not know but that it might be different : and it's present height corresponding so well with that of the *Pygmies* of the Ancients, (and we may allow something for growth too) induced me to this *denomination* : For as *A. Gellius* (88) tells us, the *Pygmies* were two Foot and a quarter high. *Pygmæos quoque* (saith he) *haud longè ab his nasci, quorum qui longissimi sunt, non longiores esse quàm pedes duos & quadrantem.* And so *Pliny* (89), *Suprà hos extremâ in parte Montium Trispithami, Pygmæique narrantur, ternas Spithamas longitudine, hoc est ternos dodrantes non excedentes* ; that is twenty seven Inches. For as *Ludovicus Vives* (90) observes, a *Foot* contains sixteen *Digiti* or twelve *Pollices*. The *Dodrans* or *Spithama*, which is the *Palmus major*, contains nine *Pollices* ; the *Palmus minor* is but three *Pollices*, or four *Digiti*, that is, a quarter of a Foot : And so *Herodotus* (91) informs us, that the *Palmus* contains four *Digiti*, and the *Cubit* six *Palmi*. The *Pygmie* therefore being *Trispithamus* or three *Spithamæ* long, was twenty seven Inches long, or as *A. Gellius* tells us, two Foot and a quarter. So our *Animal*, before Dissection measured twenty six Inches ; but in the *Sceleton*, only four and twenty Inches. Not but *Strabo* (92) out of *Megasthenes*, does mention too, the πυ͑ϊασπιθάμυς ανθρώπυς, as well as the τϱισπιθάμυς ; but these latter (he tells us) were those, that *Homer* makes to fight the *Cranes*. However it be, if our *Ape* be not the *Pygmie* of the Ancients, yet I can't but think, the *Pygmies* of the Ancients were only a sort of *Apes*, notwithstanding all the *Romances* that have been made about them. And if so, and our *Ape* be found not much to exceed the measures given, I shall think my Conjecture in giving this Name, not amiss. But of this hereafter. And to proceed :

Since the *Bones* are the main Timber-work in this Fabrick of *Animal* Bodies, by which the whole is supported, and upon their Structure, in a good measure, does depend their manner of *local motion*, we will here more particularly enquire, which may be thought the most natural way of walking in our *Pygmie*, either as a *Quadruped* or a *Biped*, for it did both upon occasion ; and we will see whether by Nature 'twas equally provided for the doing both.

Now when I observed it to go upon all four, as a *Quadruped* (as has been already remark'd) it did not place the *Palms* of the *Hands* flat to

(88) *A. Gell. Noct. Attic.* lib.9.cap.4.p.205. (89) *Plinij Nat.Hist.*lib.7.cap.2.p.m.13. (90) *Lud. Vives Comment.in D. Augustini de Civitate Dei,* lib.16.cap.8.p.m.882, (91) *Herodotus in Euterpe,* N°. 149. p. m. 448. (92) *Strabo. Geograph.* lib.15.p.m.489.

the

the Ground, but went upon it's Knuckles, or rather upon the firſt Joints of the Fingers of the Fore-hands, the ſecond and third Joints being bended or touching the Ground ; which ſeem'd to me ſo unuſual a way of *walking*, as I have not obſerved the like before in any *Animal*. And I did expect it the leſs here : becauſe the Fore-limbs being ſo very long,it might be thought, that it had the leſs need of thus raiſing the Body. And the whole weight of the Body thus lying upon theſe Joints of the *Fingers*, one would think, that they ſhould be ſoon tired in ſupporting it, and that *Nature* did not deſign it for a Conſtancy, but only upon oc-caſion, or a preſent ſhift : For if it was to be it's uſual way of walking, no doubt, for it's greater eaſe, it would place the *Palms* flat to the Ground,as all other *Animals* do the *ſole* of the *Foot*, and hereby it would be rendered better able to bear this weight.

Beſides, when it walks thus upon it's *Fingers*, the *flexure* at the Elbow will be inwards, towards the ſides of the Body, which is different from all other *Quadrupeds*, and in it's Progreſſion will be of no uſe at all, nay, will be an hinderance to it ; and it will require a great tention of the *Muſcles* to keep theſe *Fore-limbs* ſtrait ; and if they are not kept ſo, they muſt halt, and can't move ſwiftly ; which makes me diffident, that this can't be it's Natural Poſture in going ; for Nature always contrives the eaſieſt and beſt ways of *Motion*. Now in *Quadrupeds* the flection of the fore and hinder Limbs, is both the ſame way : But in a *Man* and an *Ape* (as I have before remarked from *Ariſtotle*) 'tis contrary ; or as *Pliny* expreſſes it, *Homini genua & Cubita contraria, item Urſis & Simia-rum generi, ob id minimè pernicibus.* But how *Pliny* comes to bring in the *Bear* here, I do not underſtand : for if with the *Pariſians* (93) we ſhould here underſtand by *Genua*, the *Heel-bone*, and by *Cubita* a Bone of the *Carpus* (which are often longer in *Brutes* than in *Man*) then this will be a Property not peculiar to *Bears*, but might be obſerved in other *Quadrupeds* too. I ſhould rather own it as a Miſtake in *Pliny*. Nor can I aſſent to the *Pariſians*, *That all Animals have theſe Parts turned af-ter the ſame manner, whatever* Ariſtotle *may report thereof*. I muſt con-feſs I am of *Ariſtotle's* mind, and any Body may experience it in himſelf, and obſerve the *flexure* of the *Cubit* to be different from that of the *Knee* ; and where 'tis ſo, there the Motion upon all four, will be very awkward and unnatural, and as *Pliny* obſerves, it can't be ſwift.

I ſhall here further obſerve, that in *Quadrupeds* the make of the *Tho-rax*, the ſetting on of the *Scapula*, and the *Articulation* of the *Humerus*, or Shoulder-bone, are much different from what they are in *Bipeds* : for *Quadrupeds* are narrow Cheſted, and their *Thorax* not ſo round as in a *Man*, becauſe in them the *Scapulæ* are to be placed more forward upon the Ribs, and not ſo backwards as in *Men*. And the Articulation of the

---

(93) *Vide* Their Anatomic Deſcription of a Bear in their Memoirs, *p.m.* 44.

Shoulder with the *Scapula* in *Quadrupeds* lies nearer the *Ribs* ; in Man 'tis extended farther from them. Now our *Pygmie* so exactly imitating *Humane-kind* in all these Circumstances, makes me think that *Nature* did not design it a *Quadruped*, but a *Biped*. For it had a full round *Chest* or *Thorax*, and it's *Scapulæ* placed backwards, not so forwards on the *Ribs*, and the Articulation of the *Shoulder* with the *Scapula*, stood off from the *Ribs* as it do's in *Man*. And from this very Consideration *Galen* (94) tells us, That a Man, if he would, could not walk upon all four, Δεόντως ὧν ἀνθρωπ☾. (saith he) οὐδ' εἰ βαλνθείη βαδίζειν ὅτι τῇ πτἤάρων κώλων δυωᾶτ' ἂν, ἀπημμένων αὐτῷ πόρξω τῇ θωρᾳκ☾ τῇ καντα τᾶς ὠμωπλάτας ἄρθρων. i. e. *Meritò itaque Homo ne, si volet, quidem ambulare quatuor artubus queat, quòd in ipso Scapularum Articuli longè à Thorace sint abducti.* And *Galen* all along owns, that the Structure of the *Scapula* in the *Ape*, is the same as in a *Man* ; and tells us that an *Ape* is exactly neither a *Quadruped*, nor a *Biped*, but amphibious between both. For in the same *Chapter*, speaking of the *Ape*, he saith, Τὰ ᵹ καιⁱ ὠμωπλάτας ἰ κλεῖς ἀνθρώπῳ, ἰ μάλισα πρεσοικεν, ἰ τοί γ' ὃ δέομεν☾. ἐοίκεᾳ ταύτῃ τοῖς ἀνθρώποις εἰς ὠκύτηα βαδίσεως· ἀπαμφοτερίζει τοιᾳρεῖν ἐκατέροις τοῖς ᵹένοιν, ἰ ὅτι δίπυν ὃξν ἀκριβᾶς, οὔτε πετράπυν, ἀλλὰ ἰ ὡς δίπουν χωλὰν, ὃ ᵹὸ ἀκριβᾶς ὀρθὸν εἶνᾳ δύναλᾳ, ἰ ὡς τετράπυν, ἀνάπηρόν τι ἅμα, ἰ βραδὺ, διὰ τὸ πλεῖσον ἀπήχθαι τῇ θωρᾳκ☾ αὐτὸ τὸ καιⁱ ὦμον ἄρθρον, καθάπερ εἰ ἰ τῇ ἄλλων τινὶ ζώων ἀποπαὺ ὃν τῇ θωρᾳκ☾ ὀκλὸς ἀποχωρήσειεν· i. e. *Quod verò ad Scapulas & Claves attinet, homini maximè est similis, quamquàm eâ parte homini similis esse non debebat, nam quod ad ambulationis celeritatem pertinet, simia inter genus utrumque ambigit, neque enim Bipes penitus est, neque Quadrupes ; sed quatenùs est Bipes, clauda est,non enim rectè planè stare potest ; & quatenùs est Quadrupes, mutila simul est, ac tarda, quòd Humeri articulus à Thorace plurimùm sit abductus, quemadmodùm si idem articulus in alio quopiam animante à Thorace divulsus extra secessisset.* Now altho' *Galen* tells us here, that an *Ape* can scarce stand upright ; yet in another place he declares quite the contrary ; for, saith he (95), Ἔςι δ' ὠμοιότατ☾. ἀνθρώπῳ πίθηκ☾., ὡς ἂν ςρογγύλον τε μάλις' ἔχων τὸ πρόσωπον, ἰ τὰς κυνόδονλας μικρὰς, τὸ ςέρνον πλατύ, ἰ τὰς κλεῖς μακροτέρας, ἰ ἥκισα δασύς, ἰ ὀρθὸς ἱςάᾳ καλῶς, ὡς ἰ βαδίζειν ἀμέμπτως, ἰ δεὶν ὠκέως δυωίξαᾳ. i. e. *Est autem simillima homini Simia, ut quæ rotundam præcipuè habet faciem, Dentes Caninos parvos, latum Pectus, Claviculas longiores, minimùm Pilosa, quæ recta etiam stat bellè, ut & incedere sine errore, & currere velociter possit.*

We have seen upon what accounts our *Pygmie* may be thought not to be a *Quadruped*, or that it's natural *Gression* is not on all four, and how ill it is provided to go that way. We will now enquire, Whether there is not more reason to think that *Nature* designed it to be a *Biped*, and to walk erect. And in the doing this, we may obſerve the largeneſs of the *Heel-bone* in the Foot, which being ſo much extended, ſufficiently ſecures the Body from falling backwards, as the length of the *Toes* do's it's being caſt too forwards ; and the *Arms* being ſo long, may eaſily give a poiſe either way, for the pre-ſerving the *Æquilibrium* of the Body. And it may be, this is the Reaſon why the *Pongos* hold their *hands* behind their *Necks*, when they walk erect. If we conſider the *Articulation* of the *Os Femoris* in the *Acetabulum*, there is no difference to be obſerved from a *Man*, nor indeed in any other [Circumſtance that relates to this Matter. 'Tis true, in my firſt *figure* I repreſent him as weak and feeble and bending ; for when I firſt ſaw him, he was dying ; beſides, being young, and ill, it had not that ſtrength in it's Limbs, as in time and in health, it might have acquired ; and I was willing to repre-ſent what I ſaw my ſelf. But what very much ſways with me, to think him a *Biped*, and to go erect, and that *Nature* did deſign it ſo, much more than any of the *Ape* and *Monkey*-kind beſides, was my ob-ſerving the *Peritonæum* to be entire, and not perforated or protruded in the *Groin*, as it is in *Apes* and *Dogs*, and other *Quadrupeds :* as like-wiſe, becauſe I found the *Pericardium* in our *Pygmie* to be faſtened to the *Diaphragm*, as 'tis in *Man*, and which is not ſo in *Apes* and *Mon-keys*. Both which are ſo remarkable differences, and (as I have alrea-dy remarked) ſo particularly contrived for the advantage of an *erect Poſture* of the Body, that, I think, the Inference is eaſie, and we may ſafely conclude, that *Nature* intended it a *Biped*, and hath not been wanting in any thing, in forming the *Organs*, and all Parts according-ly ; and if not altogether ſo exactly as in a *Man*, yet much more than in any other *Brute* beſides : For I own it, as my conſtant Opinion, (notwithſtanding the ill ſurmiſe and ſuggeſtion made by a forward Gentleman) that tho' our *Pygmie* has many Advantages above the reſt of it's *Species*, yet I ſtill think it but a ſort of *Ape* and a meer *Brute* ; and as the *Proverb* has it, πίθηκ@ ‍‍γ̀ ὁ πίθηκ@, κᾂν χρύσεα ἔχη σύμβολα. (96) *An Ape is an Ape, tho' finely clad.*

This *Proverb*, perhaps, might have it's riſe from ſome ſuch occaſion as *Lucian* mentions in another place ; and the Story being pleaſant, and relating to what we have been juſt now diſcourſing upon, viz. it's manner of *Motion*, we will inſert it here, and then proceed to the *Myology*. *Lucian* (97) therefore ſaith, Λέγεται ‍‍γὰ‍ βασιλεύς τις Αἰγύπτιος, πιθήκ‍ας ‍ποτὲ πυξριχίζειν διδάξαι, &c. i. e. *Fertur Ægyptius Rex quidam Simias ut tripudiarent inſtituiſſe, Animaliaque (nam admodùm ad res humanas imitandas ſunt apta) celerrimè didiciſſe, ut Perſonata ac Purpura-ta ſaltarent : eratque admodùm viſu res digna, donec Spectator quiſpiam*

---

(96) *Lucian. adverſus indoctum. Oper.* p. m. 865.   (97) *Lucian. Piſcator ſive Reviviſcentes.* p. m. 214.

urbanus

urbanus nuces è finu depromptas in medium abjiceret : id fimiæ videntes, tripudij oblitæ, id quod erat , fimiæ pro faltatoribus evaferunt , Perfonas conterebant, vefitum difcerpebant, invicemque pro fructibus depugnabant, ità ut Pyrriches ordo diffolveretur, à Theatroque ridebatur. And in another place (98) he tells the like Story of *Cleopatra's Apes.* So that they can, not only go erect, but can dance in a figure too,if taught to do fo. But this is not *natural,* but acquired by *Art* ; and even Dogs have been taught to do the fame. So *Ælian* (99) tells us, that an *Ape* is eafily taught to perform any Action ; if 'tis taught to Dance, 'twill Dance, or Play upon the Pipe ; and that once he faw one fupply the Place of a Coachman ; holding the Reins ; pulling them in, or letting them loofe, and ufing the Whip, as there was occafion. And that Story in *Kercher* (100), of the Embaffie that the *King* of *Bengal* fent to the *Great Mogul* in the Year 1660. is very remarkable, where a great *Ape* richly adorned, did drive a Chariot magnificently gilded, and fet with Jewels ; and did it with the greateft State and Pageantry in the World, and as skilfully as the beft Coach-man could do.

It would be infinite to relate all the Stories that are told us of them ; and I have been too tedious already. I fhall therefore haften now : But muft inform the Reader, that I am obliged to my good Friend Mr. *Cowper,* not only for defigning all my *figures;* but obtained of him likewife to draw up this enfuing account of the *Mufcles* ; whofe great Skill and Knowledge herein, is fufficiently made evident by his *Myotomia Reformata,* or, *New Adminiftration of all the Mufcles in Humane Bodies,* publifhed fometime fince : To which I refer my *Reader,* for a fuller account of them, whenever 'tis faid , that fuch and fuch *Mufcles* in the *Pygmie* refembled thofe in *Humane Bodies.* And for his greater Eafe, there are References all along made, to the *figures* ; where the firft Number fignifies the *Figure,* or *Table* ; the fecond Number the *Mufcle* exhibited or reprefented there.

(98) *Lucian. pro Mercede conductu,* p.m.363. (99) *Ælian. Hift. Animal.* lib.5.p.m.26. (100) *Kercher. China illuftrata,* Part.4.cap.7. p.m.195.

THE
MYOTOMY
OR
DESCRIPTION
OF THE
MUSCLES.

Of the Muſcles *of the* Abdomen.

THE *Obliquus Deſcendens* (Fig. 3. 38.) agreed in it's ſituation and progreſs, with that of a *Humane* Body, as the accurate *Galen* and *Veſalius* deſcribe it, and did not partly ſpring from any of the *Tranſverſe Proceſſes* of the *Vertebræ* of the Loins ; or their Ligaments and Membranes, as the later Writers would have it in *Humane* Bodies. Neither did any part of the *Obliquus Aſcendens* (Fig. 3. 39.) ariſe from the *Lumbal Vertebræ*, as *Veſalius* deſcribes it in *Men :* but agreed with the Deſcription of *Galen*, and did not differ from the *Humane*. *Drelincourt* obſerves the like in *Apes :* The ſame Author takes notice, that the *Pyramidales* are wanting in thoſe *Animals* ; which were abſent alſo in the *Pygmie*. The *Rectus* (Fig. 3. 40.) agreed with the *Humane*, and had no Connection with a Muſcular Portion, ſpringing either from the *Clavicula* or firſt *Rib*, as *Veſalius* has figured *Galen's* Deſcription of it in *Apes* and *Dogs*. The *Pariſians* ſay, *In Monkeys it aſcends to the top, paſſing under the* Pectoralis *and* Little Serratus *, it was Fleſhy only to the half of the* Sternum, *the reſt being but a meer Tendon.* Drelincourt obſerves the *Tendinous Inſcriptions* of theſe *Muſcles* in *Apes*, appear'd only on their inſide, and not on the out. The *Tranſverſalis* in this, as in moſt *Quadrupeds*, did not differ from that in *Man*.

The

The *Cremafter Mufcles* were very fmall by reafon of the leannefs of the Subject. The *Accelerator Spermatis* (Fig. 7. G.) *Erector Penis* (Fig. ib. K.) and *Tranfverfalis Penis* (ib. L.) agreed in their Situation and Figure with thofe of *Men*; the laft of which only varied in it's Termination, as appears in the *Figure*.

The *Detrufor Urinæ* agreed with the Figure of the *Bladder* of *Urine* of this *Animal*. The *Sphincter Veficæ* differ'd not from that in *Men*; and moft, if not all *Quadrupeds*; it being placed in the *Neck* of the *Bladder*, beyond the *Caruncula* or *Caput Gallinaginis*, immediately above the *Proftates*. The *Sphincter Ani* differ'd not from the *Humane*; unlefs it might feem fomewhat lefs. The *Levatores Ani* were longer and more divided from each other, than in *Humane* Bodies: The like may be obferv'd in moft, if not all *Quadrupeds*; by reafon of the Length and differing Figure of the *Bones*, whence thefe *Mufcles* take their rife.

I could find no *Occipital* nor *Frontal Mufcles* in this *Animal*.

The *Orbicularis Palpebrarum* (**Fig. 3. 2.**) and *Aperiens Palpebram Rectus* agreed with the *Humane*, and thofe of moft *Quadrupeds*. The *Obliquus Superior*, *Inferior*, *Elevator*, *Depreffor*, *Adductor*, and *Abductor Oculi*, agreed with thofe of the *Humane* Eye and an *Ape's*, as *Julius Cafferius Placentinus* Figures them Tab. 4. *Organi Vifûs*, Fig. XII. & XIII. Nor was there any *Mufculus Septimus Brutorum* in this *Animal*. The *Alæ Nafi* of the *Pygmie* being fmall, thofe *Mufcles* only appear'd, which from their Office are call'd *Conftrictores Alarum Nafi, ac Depreffores Labij fuperioris*.

The *Quadratus Genæ*, or *Platufma Myoides*, by reafon of the Leannefs of the Subject, (as I fufpect) did not appear Flefhy. The *Buccinator* (Fig. 3. 7.) was longer than that in *Man*. Nor was it any where intertext with various orders of Fibres, as *Anatomifts* commonly reprefent it in *Man*; or feem'd to arife from any other Parts, but the *Proceffus Coronæ*; from whence it's Fibres had a ftrait progrefs to the Angle of the *Lips*; as in *Men*: This and the former *Mufcles*, are counted *Common Mufcles* to the *Cheeks* and *Lips*.

The *Mufcles Common* to both *Lips*, are the *Zygomaticus*, (Fig. 3. 3.) *Elevator*, *Depreffor*, and *Conftrictor Labiorum*, which were not fo confpicuous, as in *Men*. The *Proper Mufcles* of the *upper* and *under* Lip, were very diftinct in this *Animal*, (*viz.*) the *Elevator* and *Depreffor Labij Superioris*, (Fig. 3. 4.) the laft of which is mentioned above, and called *Conftrictor Alæ Nafi*; the *Depreffor* and *Elevator Labij Superioris*, (Fig. 3. 5.)

Tho' *

Tho' the *Auricula* or *Outward Ear* of this *Animal* was as large, if not larger than that of a *Man*, yet I could not obſerve any *Muſcle*, which ſerv'd for it's *Motion*. I could not examine the *Muſcles* of the *Tympa-num* and *Stapes*, by reaſon the *Bones* were kept entire for a *Sceleton*.

The *Sternohyoideus*, *Coracohyoideus*, *Mylohyoideus* and *Geniohyoideus*, did not differ from thoſe in *Men*; which *Drelincourt* has alſo obſerved of the former in the *Female Ape*. The *Stylohyoideus* did not ariſe from the *Styliform Proceſs*; that *Proceſs* being wanting in this *Animal*, or at leaſt did not appear, by reaſon it was young; this *Muſcle* therefore ſeem'd to ariſe from the *Os Petroſum*.

The *Genioglossus*, by reaſon of the length of the *Lower Jaw*, was lon-ger than that in *Man*. The *Ceratoglossus* and *Styloglossus* differ'd not; except that the latter ariſes from the *Os Petroſum*, like the *Stylohyoideus*. The other *Muſcles* appear'd in this *Animal* belonging to it's *Tongue*. The *Sternothyroideus*, *Hyothyroideus*, *Cricothyroideus*, *Cricoarytænoideus*, *Poſti-cus* and *Lateralis*; the *Thyroarytænoideus*,and *Arytænoideus* varied not from thoſe in *Men*. The *Muſcles* of the *Fauces* alſo, differ'd not from thoſe in *Man*, (viz.) The *Stylopharyngæus*, *Pterygopharyngæus*, *Oeſophagæus* and *Vaginalis Gulæ*. The following *Muſcles* of the *Gargareon* were ex-actly like the *Humane*, (viz.) the *Sphenoſtaphylinus* and *Pterygoſtaphy-linus*.

Now all the *Muſcles* of the *Lower Jaw* may be ſeen without incommo-ding any hereafter mentioned. The *Temporalis* (Fig. 3. 1.) and *Maſſe-ter* (Fig. 3. 6.) ſeem'd ſomewhat larger than the *Humane*, and as they are commonly in *Brutes*, by reaſon their lower *Jaw-bones* are larger than thoſe of *Men*; yet theſe *Muſcles* were not ſo ſtrong, as thoſe of *Monkeys*, as the *Pariſians* repreſent them. The *Superior Salival Duct* paſt over the *Maſſeter*, and entred the *Muſculus Buccinator* of the *Pygmie*, as in *Man*. The *Digaſtricus* aroſe not from the *Mammiform Proceſs*, as in *Men*; but ſprang from the *Occipital-bone*; it's progreſs in this *Animal* agreed exact-ly with that in a *Humane* Body. *Drelincourt* deſcribes it in *Apes* thus, *Tendinem habet intermedium pollice longum, & gracilem, enaſcitur, autem non ab Apophyſe Styloide, ſed ab oſſe Baſilari*.

The *Muſcles* of the *Thorax* which appear on the fore-part come next. The *Intercoſtales externi* and *interni*, (Fig. 4. 32.) *Triangularis*, *Scalenus Primus*, *Secundus* and *Tertius*; *Subclavius* (Fig. 3. 34.) *Serratus minor anticus*, (Fig. 3. 35.) *Serratus major anticus*, (Fig. 3. 37.) All theſe were like the *Humane*. The *Pariſians* tell us, That the *Great Serratus* did in in their *Monkeys* ariſe from the fourth, fifth, and ſixth *Vertebra* of the *Neck*; but it was not ſo in the *Pygmie*: The like is taken notice of by *Drelincourt* in *Apes*. The *Diaphragma* was larger in this *Animal*, than in

Man,

Man, agreeable to the Capacity of its *Thorax :* The reſt of the *Muſcles* of the *Thorax* appear on it's Back-part, which we ſhall mention hereafter.

Before I paſs to the *Muſcles* on the Back-part of our *Pygmie*, I ſhall take notice of a Pair of *Muſcles*, that do not appear in *Humane Bodies*; which from their Uſe may be call'd *Elevatores Clavicularum*, (Fig. 3. 12.) Either of them ariſes Fleſhy from the *Transverſe Proceſſes* of the ſecond and third *Vertebra* of the *Neck*; and deſcends obliquely outwards to it's broad *Inſertion* at the upper part of the *Clavicula* ; when it Acts, it draws up the *Clavicle*, aſſiſting the *Elevator Scapulæ*, and upper part of the *Cucularis*, in raiſing the whole *Shoulder*. The ſituation of this *Muſcle*, is not unlike the upper part of that repreſented by *Veſalius* in his ſixth Table of the *Muſcles* O. T. P. Q. which he ſays is found in *Dogs* and *Apes*, and deſcribed by *Galen* in *Humane* Bodies , in whom *it is* not exiſtent. *Drelincourt* calls it *Levator Omoplatæ*, (adding) *ab Apophyſibus transverſis cervicalibus in Acromion & extremum claviculæ extenditur.*

The *Muſcles* imploy'd in the Motion of the *Scapula*, are the *Cucularis.* (Fig. 4. 1.1.1.) *Rhomboides* (Fig.4.6.) *Levator Scapulæ* (ib.5.) Theſe alſo agreed with the *Humane :* The like being taken notice of by *Drelincourt* of the *Cucularis*, in the *Female Ape*. The reſt of the *Muſcles* of the *Thorax*, are the *Serratus ſuperior poſticus*, (Fig. 4. 7.) the *Serratus inferior poſticus* (Fig. 4. 32. 32.) Theſe differ'd not from thoſe in *Men.* The *Sacrolumbalis* (Fig. 4. 29.) was not ſo thick as in *Men*, but was every way ſlenderer.

The *Muſcles* imploy'd in the *Motion* of the *Head* of the *Pygmie*, differed very little from thoſe in *Man*; as the *Splenius* , (Fig. 4. 2.) *Complexus*, (Fig. 4. 4.) *Rectus major*, *Rectus minor*, *Obliquus Superior*, and *Obliquus Inferior*, neither was this *Inferior Oblique Muſcle* larger than in *Man* ; as *Veſalius*, *Lib.* II. *Cap.* XXVIII. aſſures us, it is in *Apes* and *Dogs.* The *Maſtoideus* (Fig. 3. 8. 8.) was chiefly inſerted to the *Occipital-bone*, as the *Pariſians* obſerve it in *Monkeys.* The *Rectus internus major*, not commonly deſcribed by Authors in *Humane Bodies* , tho' it is very plain and conſtant in all thoſe, I have hitherto lookt for it, was alſo in the *Pygmie*. The *Rectus internus minor*, or *Muſculus Annuens* , ſometimes obſerved by me in *Humane Bodies*, was alſo in this *Animal* ; and ſo was the *Rectus Lateralis* deſcribed by *Falloppius* in *Men*. Nor was any of thoſe *Muſcles* I have diſcovered in *Humane Bodies*, wanting in this *Animal*, but the *Interſpinales Colli.*

The *Longi Colli* of this *Animal*, appear'd to be longer and larger than thoſe of *Humane* Bodies. The *Spinalis Colli* and *Transverſalis Colli* were like thoſe in *Humane* Bodies. The *Interſpinales Colli*, which I have elſewhere deſcribed in *Men*, did not appear in this *Animal*. The *Longiſſimus.*

mus Dorfi (Fig. 4. 28.) not unlike the *Sacrolumbalis* above noted, was
not fo thick and flefhy at it's Origin from the *Os Ilium* , *Sacrum* , and
Vertebræ of the Loins ; nor was it's external Surface in the *Pygmie* fo
tendinous, as in *Humane* Bodies; but was fomewhat broader.   The
Quadratus Lumborum was longer than in *Men*, agreeable to the fpace
between the *Spine* of the *Os Ilium*, and lower *Rib* of this *Animal.*   See
the Figure of the *Sceleton.*   The *Sacer*, and *Semifpinatus* , differ'd not
from the *Humane*, as I have reprefented them in my *Myotomia Reformata*,
·pag. 135·

The *Mufcles* of the Superior Parts and Trunk of the Body being di-
fpatch't,we come next to thofe of the *Limbs* ; and firft of the *Arm* or *Os
Humeri.*   The *Pectoralis* (Fig. 3. 33.) was much broader at it's Original,
from the *Sternum*, than in *Man :* it's Fibres were decuffated near it's
Infertion. · *Galen* and *Jacobus Sylvius* take notice of another *Mufcle* under the *Pectoralis* in *Apes*, which is implanted into the *Arm* near the
Pectoral Mufcle.   The *Deltoides* (Fig. 3. 15. and 4. 12.) was alfo broader at it's Original.   *Jac. Sylvius* tells us, this *Mufcle* in *Apes* is like that
of a *Man.*   The *Supraspinatus* (Fig. 4. 8.) agreed with the *Humane* in
it's fituation ; but was fomewhat broader at it's Origin from the upper
part of the *Bafis Scapulæ.*   The *Infraspinatus*, as the former *Mufcle* was
broader at it's Original from the *Scapula*, this on the contrary was there
narrower than the *Humane.*   *Sylvius* and *Drelincourt* mention thefe *Mufcles* in *Apes* ; but whether they refemble thofe of *Men*, or this *Animal* ,
do's not appear by their Accounts.   *Teres minor*, (Fig. 4. 10.) this is
fometimes wanting in *Men :* it was fomewhat fhorter and thicker in this
Animal.   The *Teres major*, (Fig. 4. 11.) was very large in the *Pygmie.*
The *Latissimus Dorfi* agreed with the *Humane* in it's Original and Progrefs towards the *Arm* ; but when it arrived at the *Axilla*, it parted with
a flefhy Portion, which defcended on the infide of the *Arm*, with the
Mufculus Biceps, and becoming a flender Tendon is inferted to the in-·
ternal protuberance of the *Os Humeri :* (*vide* Fig. 8. C.) which reprefents the production of this *Mufcle.*   This *Appendix* or *Acceffory Mufcle*
of the *Latissimus Dorfi*, is not peculiar to this *Animal* ; the like being
found in *Apes* according to *Jacobus Sylvius* , who, I am inclin'd to think
is miftaken, in reprefenting it's Infertion at the *Olecranum* of that *Animal :* This part of the *Latissimus Dorfi* feems a proper Inftrument in
turning the *Os Humeri* to a prone Pofition, when. thefe *Animals* go on
all four, for the more advantagious ftepping with the Fore-feet , by
raifing the *Os Humeri*,and turning it backwards.   *Galen* in *Lib. de Mufculis, Cap.*XIX. defcribes this *Appendix* of the *Latissimus Dorfi*, under the
Title of a *fmall Mufcle* found in the Articulation of the *Shoulder.*   The
Coracobrachialis was like that in *Man*, but had no divifion in it for any
Nerve to pafs through.   The *Subfcapularis* was alfo like that in *Man.*

The

The *Muscles* employed in Bending and Extending the *Cubit*, differ'd very little from the *Humane*, viz. *Biceps*, (Fig. 3.16.16.) *Brachiæus internus*, (ib. 18.) *Gemellus*, (Fig. 4. 14.) *Brachiæus externus*, *Anconæus*, (Fig. 4. 15. 15.) The like is observed of these *Muscles* by *Sylvius* in *Apes*, who only adds that the *Extenders* are remarkably large in that *Animal*. The *Biceps* in the *Pygmie*, had the same double tendinous Termination, as in *Man*.

The *Caro Musculosa Quadrata* appear'd in the *Palm* of the *Pygmie :* nor was there any fleshy Belly, and long Tendon to the *Palmaris* ; yet there was a *Tendon* or Ligament extended in the *Palm* ; the like has been often taken notice of in *Men*, as *Realdus Columbus* also observes. The *Parisians* tell us, the *Palmaris in Monkeys is extraordinary large.*

The *Muscles* of the *four Fingers* were, the *Perforatus*, (Fig. 3. 24.) *Perforans*, (Fig. 3. 25.) *Lumbricales* ; (ib. 31.) these agreed exactly with the *Humane* ; but the *Extensor Digitorum Communis* (Fig 4. 21.) was larger and distinct from the *Extensor minimi Digiti*, as in *Men* and *Apes*, which *Drelincourt* observes. The *Extensor Indicis*, *Abductor Indicis*, (Fig. 3. 30.) *Extensor minimi digiti*, (Fig. 4. 20.) *Abductor minimi digitis* (Fig. 4. 25.) and *Interossij Manûs*, differ'd not from those in *Men*. All the *Muscles* of the *Thumb* resembled those in *Men*, (viz.) the *Flexor tertij internodij pollicis*, *Abductor Pollicis*, (Fig. 3. 28.) *Flexor primi & secundi ossis pollicis*, (ib. 29.) *Adductor Pollicis*, (Fig. 4. 27.) *Extensor primi internodij Pollicis*, (ib. 23.) *Extensor secundi ossis Pollicis*, and *Extensor tertij ossis pollicis*. The *Muscles* of the *Wrist* also agreed with those in *Men* ; viz. the *Flexor Carpi Radialis*, (Fig. 3. 23.) and *Ulnaris*, (ib. 26.) the *Extensor Carpi Radialis*, (ib. 19.) and *Ulnaris* ; (ib. 20.) The two last *Drelincourt* says, are also like the *Humane* in the *Male-Ape*.

The *Muscles* employ'd in the *Pronation* and *Supination* of the *Radius* in the *Pygmie*, were larger in proportion than those in *Men*. The *Pronator Radij teres* (Fig. 3. 20.) had a double Origin ; the one from the internal Protuberance of the *Os Humeri*, the other from the upper part of the *Ulna :* the Pronator *Radij Quadratus*. The *Supinator Radij Longus* is taken notice of by *Drelincourt* in *Apes* to be like that of *Men*. The *Supinator Radij brevis*, (Fig. 4. 24.) agreed exactly with the *Humane*.

The *Muscles* of no part disagreed so much from those in *Men*, as those of the *Thigh* of this *Animal :* Here was no *Glutæus minor* ; nor did the *Glutæus maximus* (Fig. 4. 33. 33.) resemble the *Humane :* It was meerly *Tendinous* at it's Origin, from the whole *Spine* to the *Os Ilium* ; it was much longer, and not so thick as in *Man* ; nor were it's fleshy Fibres so divided : This *Sylvius* describes for the *Membranosus* in *Apes*. The *Parisians* give a very imperfect account of the *Musculi Glutæi* in *Monkeys*, where they tell us, *The Muscles of the Buttock had a Figure differing from*

N   *those*

thoſe in Men; being ſhorter, by reaſon the Oſſa Ilium *in Apes are much ſtraiter than in Man.* The *Glutæus medius* was alſo longer than that in *Man.*
The *Pſoas magnus* was alſo longer ; which *Sylvius* (from it's Figure I
ſuppoſe) calls *Lumbaris Biceps* in *Apes.* The *Pſoas parvus* was alſo longer and larger, than in *Man.* Beſides this, the *Pariſians* tell us of two
other little *Muſcles* in *Monkeys,* which have the ſame Origin as the *Pſoas* ;
and were inſerted into the upper and inward part of the *Os Pubis.* The
Iliacus Internus was long, conformable to the Figure of the *Os Ilium* of
this *Animal* ; *(Vide* Fig. 5. 28. 28.) The *Pectineus* was not very diſtinct.
The *Triceps* (Fig. 4. 37.) had no Tendinous Termination at the lower
Appendix of the *Thigh-bone* internally. *Jacobus Sylvius* ſays in *Apes,*
*Tricipitis pars longiſſima à Tubere in Condylum : altera portio inſignis, à Tu-
bere etiam nata, poſtico cruri propè toti affixa, ad uſque Cavitatem Inter duos
condylos mediam : tertia minima & breviſſima oſſis pubis in medium & po-
ſticum Os Cruris.* The *Pyriformis* (Fig. 4. 35.) was like the *Humane* ;
nor did it appear leſs in proportion, as the *Pariſians* repreſent it, in
Monkeys, who ſay, *This Muſcle, inſtead of taking it's riſe from the lower
and external part of the* Os Sacrum, *it proceeded from the* Iſchium *near
the* Cavitas Cotyloides. The *Marſupialis* had it's *Marſupium* much broader than in *Men.* The *Quadratus Femoris* was leſs than in *Man.* The
Obturator extrorſum was much larger.

The *Common Muſcles* of the *Thigh* and *Leg,* agreed in their Situation
and Number, with thoſe of *Men.* The *Membranoſus* (Fig. 3. 41.) had
not ſo ſtrong a Tendon to cover the *Muſcles* of the *Thighs* and *Tibia,* as
in *Man.* The *Sartorius* (Fig. 3. 42.) agreed with the *Humane.* The
Gracilis (Fig. 3. 48.) was thicker and larger near it's Origin. The *Se-
minervoſus* (Fig. 4. 40.) and *Semimembranoſus,* differ'd not from the
Humane. The *Biceps* (Fig. 4. 41.) had it's ſecond beginning, ſomewhat lower, than in *Men :* The *Pariſians* tell us, *The Biceps in Monkeys
had not a double Origin as in* Man, *but proceeded intire, from the Knob of
the* Iſchium *, and was inſerted to the upper part of the* Perona. *This ſingle
Head was in requital very thick and ſtrong.* The *Rectus* had a double order of Fibres, as in *Man.* The *Popliteus,* I muſt confeſs eſcap't my notice. *Sylvius* tells us, in *Apes,* it agrees with *Men.* The reſt of the *Muſcles* of this part, which we eſteem *Proper* to the *Tibia,* and ariſe from the
Os Femoris, were much leſs than the *Humane,* as the *Vaſtus Internus ,*
(Fig. 3. 44.) *Crureus,* and *Vaſtus externus.*

The Muſcles of the *Tarſus* or *Foot,* agreed in Number and Situation
with the *Humane* ; but varied in their Figure. The *Gaſtrocnemius externus* (Fig. 4. 43.) had not ſo large a Belly, nor were it's Fibres ſo va-
riouſly diſpoſed ; but it continued fleſhy much lower, than in *Man.*
Sylvius tells us in *Apes, Capita Gemellorum* (meaning this *Muſcle*) *Oſſa
Seſamoidea habent , firmantia in Condylis Crus cum Tibia.* The *Plantaris*
differ'd not from that in *Man.* The *Gaſtrocnemius internus,* or *ſoleus,*
(Fig.

(Fig. 4. 44.) continued fleshy to the *Os Calcis*, as *Sylvius* observed it in *Apes*. The *Tibialis Anticus* (Fig. 3. 49.) was much larger, and continued fleshy much lower, than in Man. *Sylvius* observ'd an *Os Sesamoides* in the Tendon of this *Muscle* in Apes. The *Peroneus primus* (Fig. 3. 51.) differ'd very little from that in *Man*; it's Tendon having the same progress in the Bottom of the *Foot*, to the Bone of the *Metatarsus* of the *Great Toe* ; which is nevertheless denied by *Galen* to be existent in *Man* ; for which *Vesalius, lib. 2. cap.* 59. severely Censures him. I have more than once, seen a *Boney body.*, placed in this *Tendon* at it's *Flexure* on the *Os Cuboides* in *Humane Bodies :* The like is taken notice of by *Sylvius* in an *Ape*. The *Peroneus secundus* differ'd not from that in *Man*. The *Tibialis Posticus* (Fig. 4. 45.) was not so large as in *Man*.

The *Muscles* of the *Great Toe* differ'd from the *Humane*. The *Extensor Pollicis longus* (Fig. 3. 52.) had a more Oblique progress , and was fleshy lower. The *Extensor Pollicis brevis* (Fig. 3. 53.) was much larger, and it's progress on the *Foot* almost transverse. The *Flexor Pollicis longus* was pretty large. The *Flexor Pollicis brevis* (Fig. 4. 47.) was very large, and inseparably joined with the *Abductor* , which was very little. The *Parisians* tell us, *The Great Toes of the Monkeys had Muscles like those of a Man's Thumb.* The *Extensor Digitorum Pedis longus* (Fig. 3.53.) had no Tendon implanted on the *Os Metatarsi* of the *Little Toe*. The *Perforatus* (Fig. 4. 46.) *Perforans*, (ib. 48.) *Lumbricales*, and *Abductor minimi Digiti*, differ'd very little from those in Men. The *Musculus Extensor Digitorum brevis*, and *Transversalis Pedis* did not appear in this *Animal*.

I shall not at present give the *Reader* the trouble of the *Reflections*, that I intended, upon the Observations made in the *Anatomy* of this remarkable Creature ; since I am conscious (having been so tedious already) that 'twill but farther tire him, and my self too. I shall therefore now conclude this *Discourse*, with a brief Recapitulation of the Instances I have given, wherein our *Pygmie*, more resembled the *Humane kind*, than *Apes* and *Monkeys* do : As likewise sum up those, wherein it differ'd from a *Man*, and imitated the *Ape-kind*. The *Catalogues* of both are so large, that they sufficiently evince, That our *Pygmie* is no *Man*, nor yet the *Common Ape* ; but a sort of *Animal* between both ; and tho' a *Biped*, yet of the *Quadrumanus-kind* ; tho' some *Men* too, have been observed to use their *Feet* like *Hands*, as I have seen several.

The

The Orang-Outang *or* Pygmie *more reſembled a* Man, *than* Apes *and* Monkeys *do.*

1. IN having the *Hair* of the *Shoulder* tending downwards; and that of the *Arm*, upwards.

2. In the *Face* 'twas liker a *Man* ; having the *Forehead* larger, and the *Roſtrum* or *Chin* ſhorter.

3. In the *outward Ear* likewiſe; except as to it's *Cartilage*, which was thinner as in *Apes*.

4. In the *Fingers* ; which were much thicker than in *Apes*.

5. In being in all reſpeċts deſigned by *Nature*, to walk ereċt; where-as *Apes* and *Monkeys* want a great many Advantages to do ſo.

6. The *Nates* or *Buttocks* larger than in the *Ape-kind*.

7. It had *Calves* in it's *Legs*.

8. The *Shoulders* and *Breaſt* were more ſpread.

9. The *Heel* was longer.

10. The *Membrana Adipoſa* placed here, next to the *Skin*.

11. The *Peritonæum* in the *Groin* entire ; and not perforated, or pro-truded, as in *Apes* and *Monkeys*.

12. The *Inteſtines* or *Guts* much longer.

13. The *Inteſtines* being very different in their bigneſs , or largeneſs of their *Canalis*.

14. In having a *Cæcum* or *Appendicula Vermiformis*, which *Apes* and *Monkeys* have not: and in not having the beginning of the *Colon* ſo pro-jeċted or extended, as *Apes* and *Monkeys* have.

15. The Inſertion of the *Duċtus Bilarius* and the *Duċtus Pancreaticus* in a *Man*, the *Pygmie*, and an *Ape* was at the ſame *Orifice*. In a *Monkey* there was two Inches diſtance.

16. The *Colon* was here longer.

17. The *Liver* not divided into *Lobes*, as in *Apes* and *Monkeys* ; but entire, as in *Man*.

18. The *Biliary Veſſels*, the ſame as in *Man*.

19. The *Spleen* the ſame.

20. The *Pancreas* the ſame.

21. The Number of the *Lobes* of the *Lungs*, the ſame as a *Man's*.

22. The *Pericardium* faſtened to the *Diaphragm*, as in *Man* ; but is not ſo in *Apes* and *Monkeys*.

23. The *Cone* of the *Heart*, not ſo pointed, as in *Apes*.

24. It had not thoſe *Pouches* in the *Chaps*, as *Apes* and *Monkeys* have.

25. The *Brain* was abundantly larger than in *Apes* ; and all it's Parts exaċtly formed like the *Humane Brain*.

26. The *Cranium* more globous ; and twice as big as an *Ape's* or *Monkey's*.

27. All

27. All the *Sutures* here, like the *Humane* : And in the *Lambdoidal Suture* were the *Offa triquetra Wormiana.* In *Apes* and *Monkeys* 'tis otherwife.

28. It had an *Os Cribriforme*, and the *Crifta Galli* ; which *Monkeys* have not.

29. The *Sella Equina* here, the fame as in *Man* ; in the *Ape-kind* 'tis more rifing and eminent.

30. The *Proceffus Pterygoides*, as in *Man :* In *Apes* and *Monkeys* they are wanting.

31. The *Offa Bregmatis* and *Temporum* here the fame as in *Man.* In *Monkeys* they are different.

32. The *Os Zygomaticum* in the *Pygmie* was fmall ; in the *Monkey* and *Apes* 'tis bigger.

33. The Shape of the *Teeth* more refembled the *Humane*, efpecially the *Dentes Canini* and *Molares.*

34. The *Tranfverfe Apophyfes* of the *Vertebra* of the *Neck*,and the Sixth and Seventh *Vertebra*, were liker the *Humane*, than thefe Parts in *Apes* and *Monkeys* are.

35. The *Vertebra* of the *Neck* had not thofe *Foramina* for tranfmitting the *Nerves* ; which *Apes* have and *Man* has not.

36. The *Vertebra* of the *Back*, and their *Apophyfes Recta* like the *Humane :* and in the lower *Vertebra* but two *Apophyfes inferna* ; not four, as in *Apes.*

37. There were but five *Vertebra* of the *Loins* here, as in *Man :* in *Apes* and *Monkeys* there are fix.

38. The *Spines* of the *Lumbal Vertebra* ftrait, as in *Man.*

39. The *Os Sacrum* was compofed of five *Vertebra*,as in *Man :* in *Apes* and *Monkeys* there are but three *Vertebra.*

40. The *Os Coccygis* had but four *Bones*, and thefe not perforated, as 'tis in *Man :* In *Monkeys* there are more *Bones*, and they are perforated.

41. In the *Pygmie* there were but feven *Cofta vera* ; and the Extreams of the *Notha* were Cartilaginous ; and the *Ribs* were articulated to the body of the *Vertebra.* In *Apes* and *Monkeys* there are eight *Cofta vera* ; and the Extreams of the *Notha* are *offious* ; and the *Articulation* is in the Interftices of the *Vertebra.*

42. The *Os Sterni* in the *Pygmie* was broad , as in a *Man :* in the *Monkey* 'tis narrow.

43. The *Bones* of the four *Fingers* much larger than in the *Ape-kind.*

44. The *Thigh-bone* in it's *Articulation*, and all other refpects, like the *Humane.*

45. The *Patella* round, not long ; fingle, not double ; as 'tis faid to be in *Apes.*

46. In the *Heel*, the *Tarfus*, and *Metatarfus* , the *Pygmie* was like a *Man.*

47. The *middle Toe* in the *Pygmie* was not the longeft, as 'tis in the *Ape-kind.*

48. Thefe

48. Theſe *Muſcles*, *viz*. The *Obliquus Inferior Capitis*, the *Pyriformis* and *Biceps Femoris*, were like the *Humane* ; whereas the ſame in *Apes* and *Monkeys* are different.   And Note, That all the other *Muſcles* that are not otherwiſe ſpecified in the following *Catalogue*, were like the *Humane* alſo ; but whether all the ſame *Muſcles* in *Apes* and *Monkeys* reſemble the *Humane*, could not be determined, for want of a Subjeſt to compare them with, or Obſervations made by others.

The Orang-Outang *or* Pygmie *differ'd from a* Man, *and reſembled more the* Ape *and* Monkey-kind.

1. IN the littleneſs of it's Stature.
    2. In the flatneſs of the *Noſe*, and the ſlit in the *Alæ Narium*.
3. In having a riſing Ridge of the *Cranium* under the *Eye-brows*.
4. In being more hairy behind, than before.
5. In having the *Thumb* ſo little, tho' larger than in the *Ape-kind*.
6. In having the *Palm* of the *Hand* longer and narrower.
7. In the length of the *Toes*.
8. In having the *Great Toe* ſet at a diſtance from the other, like a *Thumb*; and being *Quadrumanus*, like the *Ape-kind*.
9. In having the *Shoulder* and *Thigh* ſhorter.
10. In having the *Arms* longer.
11. In having no pendulous *Scrotum*.
12. In the largeneſs of the *Omentum*.
13. The *Gall-Bladder* long and ſlender.
14. The *Kidneys* rounder than in *Men*; and the *Tubuli Urinarij* different.
15. The *Bladder* of *Urine* longer.
16. In having no *Frænum* to the *Præputium*.
17. In having the *Bony Orbit* of the *Eye* ſo much protruded inwards, towards the Brain.
18. It had not thoſe two Cavities under the *Sella Turcica*, as in .*Man*.
19. The *Proceſſus Maſtoides* and *Styloides* very ſmall, almoſt wanting.
20. The *Bones* of the *Noſe* flat.
21. In the Number of the *Teeth*, it reſembled the *Ape-kind*.
22. The *Vertebræ* of the *Neck* ſhort as in the *Ape-kind*, and flat before, not round ; and their *Spines*, not *bifide*, as in *Man*.
23. In the firſt *Vertebra* of the *Neck* there was no *Spine*.
24. In an *Ape* the Tenth *Vertebra* of the *Back*; in a *Man* the Twelfth ; in the *Pygmie* the Thirteenth *Vertebra*, *infra ſuprave ſuſcipitur*.

25. The

Fig. 1.

25. The *Os Sacrum* altogether like the *Ape-kind*, only in the number of the *Vertebræ*.

26. In having Thirteen *Ribs* on a fide: a *Man* has but Twelve.

27. The *Bone* of the *Thumb* but fmall.

28. The *Os Ilium* perfectly like the *Ape-kind* ; being longer, narrower, and not fo Concave as in *Man*.

29. The *Bones* of the *Toes* in their length, and the *Great Toe* in it's Structure imitated the *Ape-kind*.

30. Thefe *Mufcles* were wanting in the *Pygmie*, which are always found in *Men* ; viz. *Occipitales, Frontales, Dilatatores Alarum Nafi*, feu *Elevatores Labij Superioris, Interfpinales Colli, Glutæi minimi, Extenfor Digitorum Pedis brevis*, and *Tranfverfalis Pedis*.

31. Thefe *Mufcles* did not appear in the *Pygmie* , and are fometimes wanting too in *Humane Bodies* ; viz. *Pyramidales* ; *Caro mufculofa Quadrata* ; the long Tendon and flefhy Belly of the *Palmaris* ; *Attollens Auriculam* ; and *Retrahens Auriculam*.

32. The *Elevatores Claviculariũ* are in the *Pygmie* and the *Ape-kind*, and not in *Man*.

33. Thefe *Mufcles* refembled thofe in *Apes* and *Monkeys*, and differ'd from the *Humane*, viz. *Longus Colli, Pectoralis, Latiffimus Dorfi, Glutæus maximus & medius, Pfoas magnus & parvus, Iliacus internus* , and the *Gafterocnemius internus*.

34. Thefe *Mufcles* differ'd likewife from the *Humane*, viz. the *Deltoides* ; the *Pronator Radij teres* ; the *Extenfor Pollicis brevis*.

The Explanation of the Figures.

Figure the Firft

REprefents the Fore-parts of the *Orang-Outang* or *Pygmie* , in an Erect Pofture: Where you may obferve the largenefs of the *Head* ; and broadnefs of the *Forehead* ; the jutting out of the *Eye-brows* ; the *Eyes* fomewhat funk ; the *Nofe* flat ; the *Face* without hair and wrinkled ; the *Teeth* like the *Humane* ; the *Chin* fhort ; the *Ears* ftanding off from the Head ; the *Head* hairy ; the *Shoulders* fpread and large ; the *Arms* and *Palms* of the *Hands* long ; the *Nails* like thofe in a *Man* ; the *Hair* of the *Shoulder* inclining downwards, and that on the *Arms*, upwards ; the *Fingers* large ; the *Thumb* little ; the *Breaft* full chefted and fpread ; the *Mamma* or *Teats* placed as in *Man* ; the *Belly* flat ; the *Navil* as in *Man* ; the *Penis* half-way covered with the *Prepuce*, which had no *Frænum* ; no pendulous *Scrotum* here ; the *Thighs* a little divaricated ; the

Legs

Legs long and with *Calves* ; the *Foot* like a *Hand*, having long *Toes*, and the *Great Toe* placed at a diftance from the others, like a *Thumb* ; the *Feet*, *Hands*, *Face*, *Ears*, and *Penis* without *Hair* ; and all the Fore-parts of the Body rather lefs hairy than here reprefented ; and the *Head* is too much fhrunk down between the Shoulders.

The Second Figure

REprefents the Hinder Parts of the *Pygmie* in an Erect Pofture like-wife ; where may be obferved the *Globous Figure* of the *Head* ; the ftraitnefs of the *Back* ; and that 'tis more hairy behind, than before ; the *Fingers* of the *right Hand* are reprefented bending , to fhew the *Action*, when it goes on all four ; for then it places only the *Knuckles*, not the *Palms* of the *Hands* to the Ground. The *Sole* of the *left Foot*, by rea-fon of the length of the *Toes*, and the fetting on of the *Great Toe*, looks like the *Palm* of the *Hand* : but the *right*, having fo long a *Heel*, and its *Toes* being hid, appears rather like a *Foot*, and upon occafion per-forms the Office of both, either of a *Foot* or *Hand*. A little above the *Anus*, there is a black Spot, which reprefents a fmall Protuberance of the *Os Coccygis*.

The Third Figure

REprefents the *Mufcles* which appear on the Fore-part of the *Body*.
A. Part of the Coronary Suture.
B. The Divifion of the *Cranium* made by the Saw.
c. The *Meatus Auditorius*.
d. Part of the *Os Jugale*, or *Zygomaticum*.
e. The *Parotid Gland.* * The *Salival Duct.*
f. The *Inferior Maxillary Gland*.
g. g. The *Clavicula*.
h. Part of the *Spina Scapulæ*, as joyned to the *Clavicle*.
j. The Nerves, and Blood Veffels which pafs to the Arm.
k. The Trunk of the *Nerve* in the left Arm, that goes to the Fingers.
l. A large Trunk of the *Arterie*, and a *Nerve* in the *Cubit*, as in *Hu-mane* Bodies.
m. m. The Internal Protuberances of the *Os Humeri*.
n. The *Radius* of the left Arm made bare.
o. The *Umbilicus*, or Navil.
p. The *Linea Alba*.
q. q. The Tendons of the Oblique Mufcles, call'd *Linea Semilunaris*.
r. r. The

Fig: 2.

M. Vander yacht Sculp.

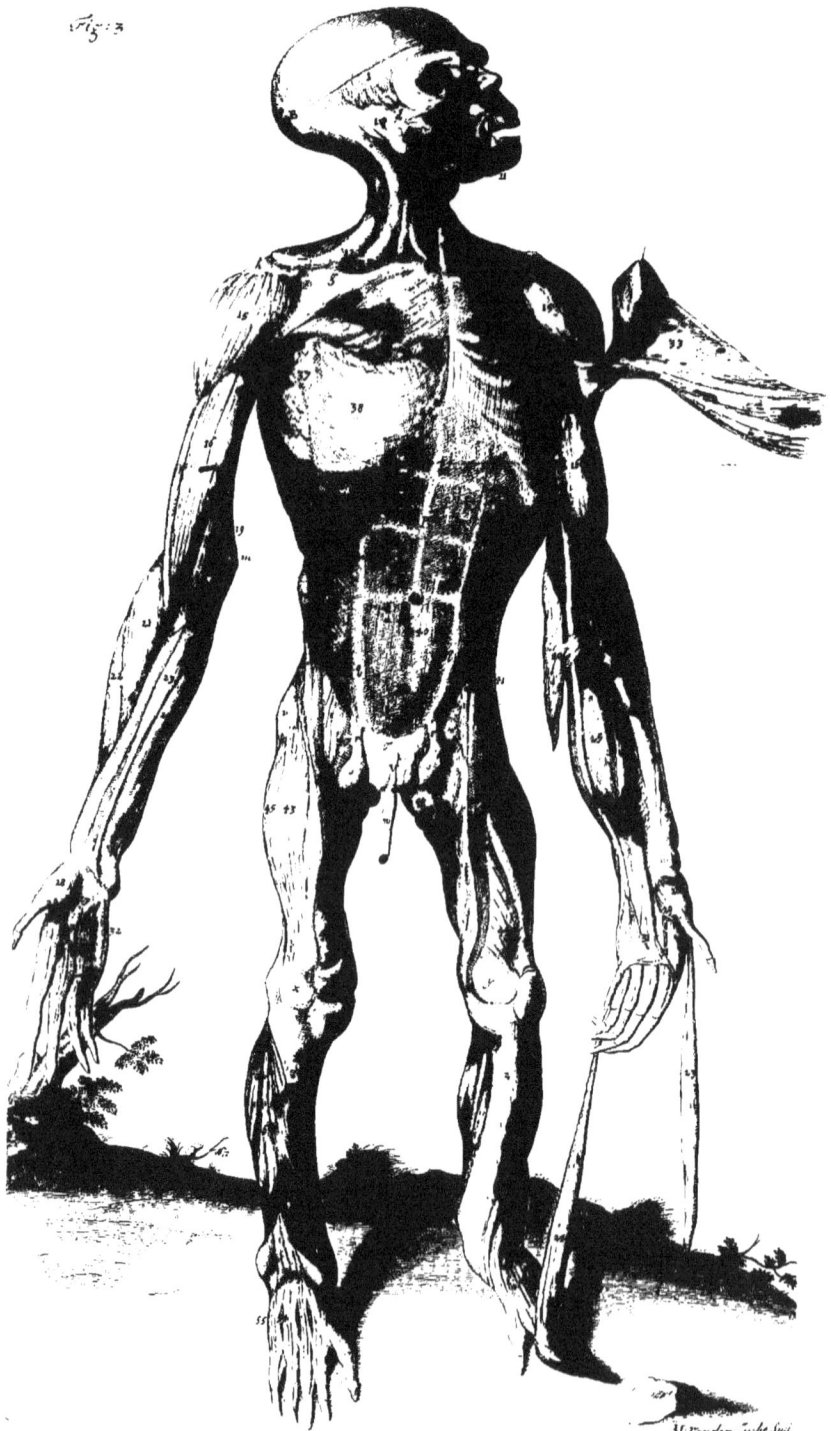

M.^r vander Gucht Sculp.

r. r. The *Tunica Vaginalis,* containing the *Vasa Præparantia,* &c.
s. s. The *Testes* or *Stones.*
t. The Blood Veſſels of the Thigh, as they paſs under the *Inguinal Glands.*
T. The *Os pubis.*
V. The *Ligamentum ſuſpenſorium Penis.*
u. The *Great Trochanter.*
w. The *Penis.*
x. x. The two *Patellæ.*
y. y. The internal and lower *Appendix* of the *Os Femoris.*
z. z. The *Tibia.*
N° 1. The *Muſculus Temporalis.*
2. The *Orbicularis Palpebrarum.*
3. *Zygomaticus, ſeu diſtortor oris.*
4. *Elevator Labij ſuperioris proprius.*
5. *Elevator Labij inferioris proprius.*
6. *Maſſeter.*
7. *Buccinator.*
8. 8. *Maſtoideus.*
9. *Sternohyoideus.*
X. Part of the *Corocohyoideus.*
11. Part of the *Digaſtricus,* and it's Inſertion into the *Chin.*
12. *Elevator Claviculæ,* which Muſcle is not in Man, but in the *Pygmie* and *Apes.*
13. Part of the *Complexus Capitis.*
14. Part of the *Cucularis.*
15. 15. *Deltoides.*
15. 16. The *Biceps.*
17. The thin Tendinous Expanſion of the *Biceps,* which involves the *Muſcles* of the *Cubit,* as in Man.
18. Part of the *Brachæus internus.*
19. The Tendinous Elongation of the *Latiſſimus Dorſi,* which is found in the *Pygmie,* and in *Apes* and *Monkeys* ; and not in Man ; near it's Inſertion into the Internal Protuberance of the *Os Humeri.*
20. 20. The *Pronator Radij teres.* That of the left ſide, being diſſected from it's Inſertion, and left at it's two Originals.
21. 21. The *Supinator Radij Longus.*
22. Part of the *Extenſor Radialis.*
23. 23. The *Flexor Radialis,* that of the left Arm hanging at it's Inſertion.
24. 24. The *Perforatus* ; that of the left ſide hanging by its Tendons on the Palm of the Hand.
25. The *Perforans* ; a little raiſed in the left Arm.
26. The Tendon of the *Flexor Ulnaris* as it runs to the *Carpus.*
27. A Tendinous Expanſion, like the *Palmaris* in Man ; but here was no Muſcle, which is often ſeen in *Humane* Bodies.
28. 28. The *Abductor Pollicis.*

O   29. The

29. The *Flexor Secundi internodij Pollicis.*
30. *Abductor Indicis,*
31. 31. The *Lumbricales.*
32. The *Abductor minimi digiti.*
33. The *Pectoralis* ; that of the left ſide being raiſed , to ſhew the de-cuſſation of it's Fibres, as in Man.
34. Part of the *Muſculus ſubclavius.*
35. *Serratus minor anticus.*
36. 36. The *Intercoſtales externi.*
37. 37. The *Serratus major anticus* ; where 'tis indented with the *Muſculus obliquus deſcendens.*
38. 38. The *Obliquus deſcendens.*
39. The *Obliquus aſcendens,* as it appeared after the *deſcendens* was re-moved.
40.40. The *Muſculi Recti,*with their *Paragraphs* or *Inſcriptions,*as in Man
41. 41. The *Muſculus communis* Membranoſi.
42. 42. The *Sartorius.*
43. 43. The *Rectus Femoris.*
44. 44. The *Vaſtus internus.*
45. Part of the *Vaſtus externus.*
46. 46. Parts of the *Triceps.*
47. 47. The *Pectinæus.*
48. The *Gracilis.*
49. 49. The *Tibialis Anticus.*
50. Part of the *Gaſterocnemius.*
51. Parts of the *Peronei.*
52. The *Extenſor Pollicis longus.*
53. The *Extenſor Pollicis brevis,* which differ'd in this *Animal,* from that in M*an.*
54. The Tendons of the *Extenſor Communis digitorum,* as they paſs be-tween the *Interoſſii.*
55. The *Abductor minimi digiti.*
56. The *Pronator Radij Quadratus.*
57. Part of the *Supinator Radij brevis* ; at it's Inſertion to the *Radius.*

Figure the Fourth

Shews the *Muſcles* of the *Back-part* of the *Body.*

a. THE *Sagittal Suture.*
        b. The *Lambdoidal Suture.*
c. c .c. The *Spines* of the *Superior Vertebræ* of the *Thorax,* and of one of th e*Inferior* of the *Neck.*

                                          d. The

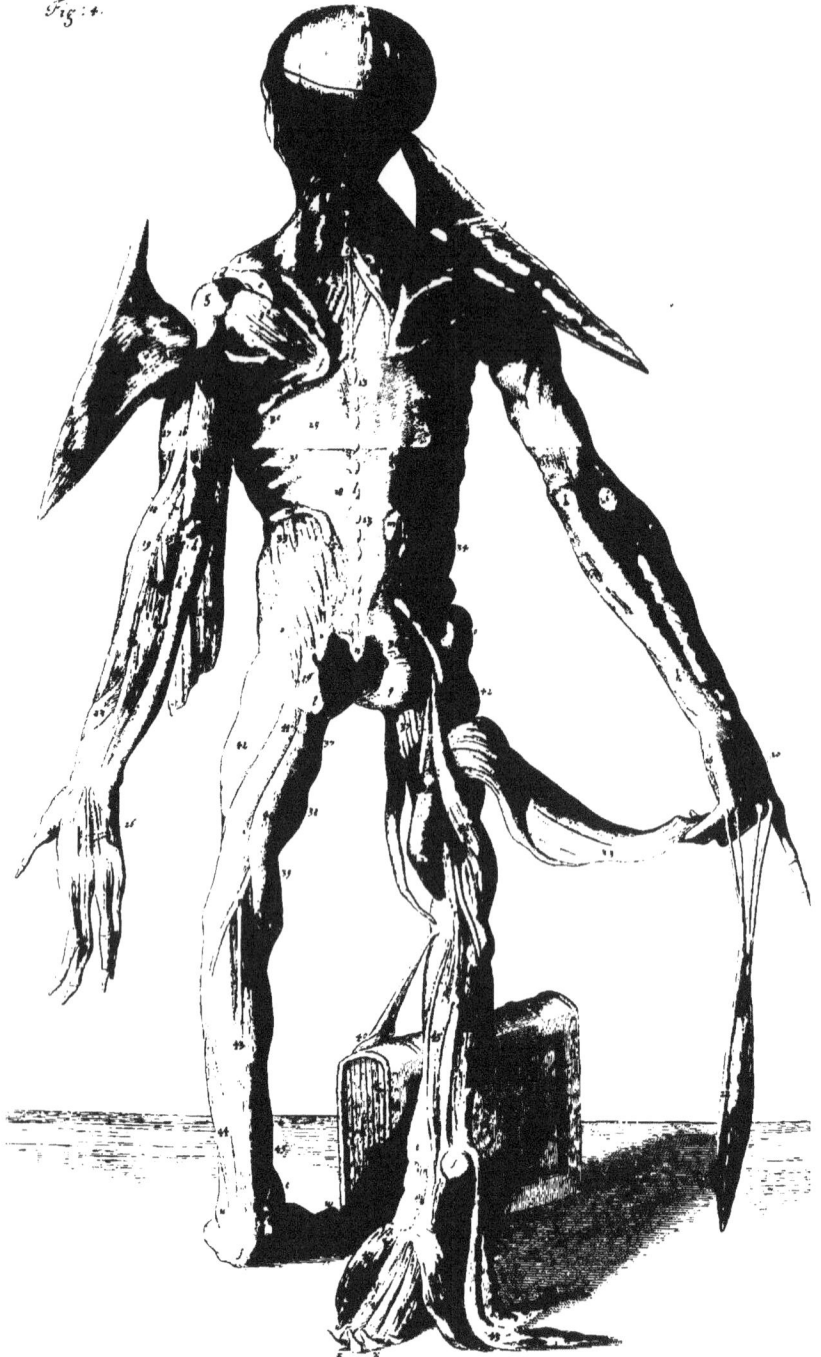

Fig: 4.

M. Vander Sucht Sculp.

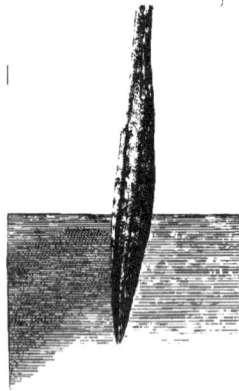

M. Vander Gucht Scul:

d. The Extremity of the *Clavicle,* where it is connected to the *Spine* of the *Scapula.*

e. The *Spine* of the *Scapula.*

f. The lower Angle of the *Scapula.*

g. The upper part of the *Os Humeri,* made bare, by raising the *Deltoid Muscle.*

h. h. The *Acromion* or *Elbow.*

i. The External Protuberance of the *Os Humeri,* where the upper part of the *Radius,* is Articulated.

k. The *Ulna.*

l. l. The *Spines* of the *Back* and *Loins.*

m. m. The *Spines* of the *Ossa Ilium.*

n. The *Os Coccygis.*

o. The *Great Trochanter.*

p. The Trunk of the Great *Crural Nerve.*

q. q. The *Ossa Ischij.*

r. r. r. The *Crural Nerves* in the Hams.

s. s. The *Os Calcis.*

t. t. The *Malleolus Internus.*

u. The *Malleolus externus.*

w. w. The *Great Toe.*

x. x. The four *little Toes.*

y. y. The *Pelvis* left open, by taking out the *Anus* with the *Rectum.*

N° 1. 1. 1. 1. The *Musculus Cucularis,* raised on the right side, and left fastened to the *Occiput,* and to its Insertion at the *Spine* of the *Scapulæ* and *Clavicle.*

2. Part of the *Splenius.*

3. Part of the *Mastoideus.*

4. Part of the *Complexus.*

5. Part of the *Levator Scapulæ.*

6. *Rhomboides.*

7. Part of the *Serratus superior posticus.*

8. *Supraspinatus.*

9. *Infraspinatus.*

X. The *Teres minor,* which is larger here than in *Man.*

11. The *Teres major.*

12. The *Deltoides* raised.

13. 13. 13. 13. The *Latissimus Dorsi,* on the right side *in situ,* in the left, freed from it's Original and hanging down.

14. The *Biceps Externus seu Gemellus.*

15. The *Anchonæus.*

16. Part of the *Brachæus internus.*

17. Part of the *Biceps internus.*

18. The *Supinator Radij longus.*

19. The *Extensor Carpi Radialis.*

O 2

20. 20. The

20. 20. The *Extenfor Carpi Ulnaris.*
21. 21. The *Extenfor Communis digitorum* , on the right fide hanging by its Tendons.
22. 22. The *Extenfor minimi digiti*, on the right fide hanging down.
23. The *Extenfores Pollicis.*
24. The *Supinator Radij brevis.*
25. The *Abductor minimi digiti.*
26. The *Mufculi interoffei.*
27. The *Abductor Pollicis.*
28. The *Longiffimus Dorfi.*
29. The *Sacrolumbalis.*
30. 30. The *Intercoftales.*
31. Part of the *Serratus major anticus.*
32. The *Serratus inferior pofticus.*
33. The *Glutæus maximus* on the left fide *in fitu* , on the right freed from its Origin, and left at it's Infertion.
34. The *Glutæus medius.*
35. The *Pyriformis.*
36. The *Marfupialis f. Obturator.*
37. 37. Part of the *Triceps.*
38. 38. The *Gracilis.*
39. The *Semimembranofus.*
40. The *Seminervofus.*
41. The *Biceps femoris.*
42. Part of the *Vaftus externus.*
43. 43. The *Gafterocnemius externus*, that of the right fide hanging to its Infertion, at the *Os Calcis.*
44. The *Gafterocnemius Internus.*
45. Part of the *flexor Digitorum perforans.*
46. The flefhy part of the *flexor Digitorum perforatus.*
47. The *flexor Offis Pollicis*, together with the *Abductor Pollicis*, raifed from it's Origin, and hanging down.
48. The *Mufculi Lumbricales.*

The fifth Figure

Reprefents the *Sceleton*, or the *Bones.*

1. THE *Os Frontis.*
2. The *Os Bregmatis.*
3. Part of the *Os Occipitis.*
4. *Os Temporale* , *feu Squammofum.*
5. *Os Jugale*, *feu Zygomaticum.*
6. The firft Bone of the *Upper Jaw.*
7. The *Os Lachrymale.*

8. The

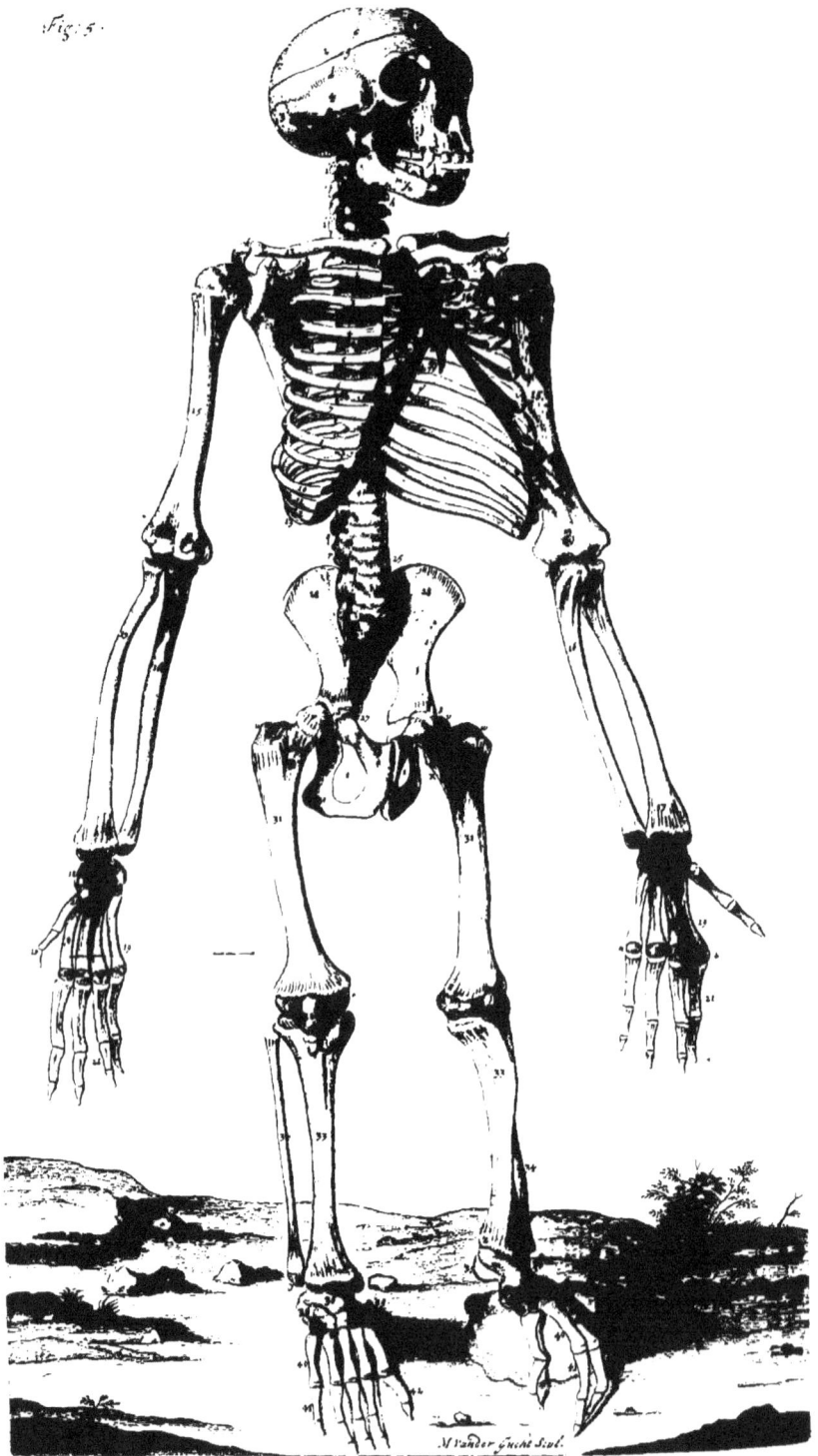

Fig: 5.

M Vander Gucht Sculp.

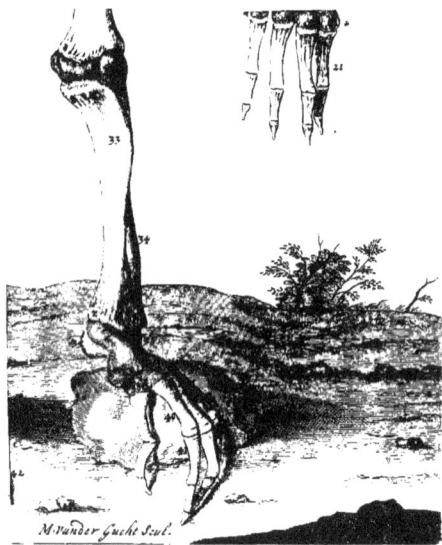

M. Vander Gucht Sculp.

8. The *Os Narium*.

9. The fourth Bone of the *Upper Jaw*.

10. The upper part of the *Os Sphænoides*.

11. The *lower Jaw*.

a. The *Proceſſus Condyloides* of the *lower Jaw*.

b. The *Proceſſus Corone*.

c. The *Coronal Suture*.

d. The *Sutura Oſſis Temporalis*, *ſeu Squammoſi*.

e. A *Foramen* for the paſſing the Nerves, and the Blood Veſſels in the upper *Jaw*.

f. A like *Foramen* in the *under Jaw*.

g. Where the *Skull* was ſawed, to take out the *Brain*.

h. The *Tranſverſe Proceſſes* of the *Vertebræ* of the *Neck*.

j. j. The *Oblique aſcending* and *deſcending Proceſſes* of the *Neck*.

12. 12. The *Vertebræ* of the *Neck*.

13. 13. The *Claviculæ*, or *Collar Bones*.

K. K. The Connection of the *Claviculæ*, to the *Spina Scapulæ*.

14. 14. The Internal parts of the *Scapula*.

l. l. The *Proceſſus Chorocoides Scapulæ*.

15. 15. The *Os Humeri*.

†. †. A *Sinus* for receiving the External Tendon of the head of the *Biceps*.

m. m. A *Sinus* for receiving the Prominence (*n.n.*) of the *Ulna* upon bending the *Arm*.

16. 16. The *Ulna*.

o. Part of the *Olecranon* of the *Ulna* of the left *Arm*.

17. 17. The *Radius*.

p. A Prominence of the *Radius*, to which the internal great Tendon of the *Muſculus Biceps* is inſerted.

18. 18. The Bones of the *Carpus*, which in a great meaſure were *Cartilaginous*.

19. 19. The Bones of the *Metacarpus*.

20. 20. The Bones of the *Thumb*.

21. 21. The Bones of the *Fingers*.

22. 22. The *Sternum* or *Os Pectoris*.

23. The *Cartilago Enſiformis*.

1. 2. 3. 4. 5. 6. 7. 8. 9. 10. 11. 12. 13. The Thirteen *Ribs* of each ſide.

24. The *Vertebræ* of the *Back*.

25. The *Vertebræ* of the *Loins*.

q. The *Tranſverſe Proceſſes* of the *Vertebræ* of the *Loins*.

r. The *Foramina* for the paſſage of the Nerves.

26. The *Os Sacrum*.

27. The *Os Coccygis*.

28. 28. The *Os Ilium*.

29. The *Os Pubis*.

30. The *Os Iſchij*.

s. s. The *Cartilaginous* Conjunction of the *Os Ilium* with the *Os Pubis* and *Iſchij* at the *Acetabulum*.   t. t. The *t. t.* The large *Foramen* of the *Os Pubis* and *Iſchij.*

31. 31. The *Os femoris.*

v. v. The Head of the *Os femoris* in the *Acetabulum.*

w. w. The *Great Trochanter,* which was *Cartilaginous.*

X. The *leſſer Trochanter.*

32. 32. The *Patella* , which was *Cartilaginous.*

33. 33. The *Tibia.*

34. 34. The *Fibula.*

35. 35. The *Os Calcis.*

36. 36. The *Aſtragalus.*

37. The *Os Cubiforme.* .

38. The *Os Naviculare, ſeu Cuneiforme majus.*

39. The *Oſſa Cuneiformia minora.*

40. 40. The *Oſſa Metatarſi.*

41. 41. The *Oſſa Digitorum.*

42. 42. The Bones of the *Great Toe.*

y. The *Malleolus externus.*

z. The *Malleolus internus.*

* * * * Signifie, that thoſe Parts were *Cartilaginous.*

The ſixth Figure

Repreſents the *Stomach, Inteſtines, Pancreas, Spleen, Liver,* &c.'

A. A. THE back ſide of the *Stomach,* it being turned upwards.

B. Part of the *Oeſophagus.* or *Gullet,* before it joyns with the upper or left *Orifice* of the *Stomach.*

C. The right *Orifice* of the *Stomach,* or *Pylorus.*

a. a. The Extremities of the *Vaſa Brevia,* which paſs between the *Spleen* and the *Stomach.*

b. b. &c. Divers *Lymphatick Glands* on the *Stomach.*

D. The *Superior Coronary Arteries* and *Veins,* and their Ramifications.

E. E. The *Inferior Coronary Blood Veſſels* of the *Stomach,* which ſends Branches alſo to the *Omentum.*

F. F. The *Omentum* or *Caul* turned up, to ſhew it's lower *Leaf.*

G. G. The *Liver,* like the *Humane ;* and not divided into *Lobes,* as 'tis in *Apes.*

c. A ſmall Lobe of the *Liver* at the entrance of the *Vena Porta.* .

d The *Fiſſure* or Cleft in the *Liver* at the entrance of the *Umbilical Vein.*

f. f. The *Gall Bladder.*

H. The beginning of the *Duodenum.*

I. I. The *Pancreas.*

g. g. The

Fig: 7.

Fig: 8.

M.r Vander Gucht Sculp.

g. g. The Blood Veſſels of the *Spleen,* eſpecially a Branch of the *Vena Porta.*

K. K. The *Spleen.*

L. L. L. The *Small Guts.*

M. The *Ileon* juſt before it enters the *Colon.*

N. The beginning of the *Colon.*

h. h. One of the Ligaments of the *Colon.*

O. O. The *Cæcum,* or *Appendicula Vermiformis.*

P. P. The *Colon* in its whole Progreſs, to the *Rectum.*

j. j. Part of the *Meſenterie.*

k. k. The Glands of the *Meſenterie.*

l. That part of the *Meſenterie,* which is connected to the *Cæcum,* or the *Meſocæcum.*

m. The *Meſocolon,* or that Part of the *Meſenterie* that is faſtened to the *Colon.*

Q. The upper part of the *Inteſtinum Rectum.*

The ſeventh Figure

Shews the *Organs* of GENERATION.

A. THE back part of the *Bladder* of *Urine,* the greateſt part of the *Bladder* being cut off.

B. The *Penis.*

C. C. The two *Ureters.*

D. D. The *Vaſa Deferentia.*

E. E. The *Veſiculæ Seminales.*

F. The *Glandulæ Proſtatæ,* or *Corpus Glandoſum.*

G. The *Bulb* of the *Cavernous Body* of the *Urethra,* covered with the *Muſculus accelerator Urinæ ſeu Spermatis.*

a. a. The two Productions of the laſt mentioned *Muſcle,* which are inſerted to the two *Cavernous Bodies* of the *Penis,* on each ſide the *Urethra,* by which means that part of the *Urethra* is compreſt, and it's Contents forced out.

b. b. The beginning of the two *Cavernous Bodies* of the *Penis.*

H. The Cavernous Body of the *Urethra.*

j. One of the *Tranſverſe Muſcles* of the *Penis,* call'd the *Third Pair.*

K. K. The *Muſculi Directores Penis.*

The

The Eighth Figure

Exhibits part of the *Musculus Latissimus Dorsi* dissected.

A. A. **T**HAT part of the *Muscle* that lies on the *Back*, as in *Humane Bodies.*

B. It's *Tendon* which is inserted to the *Os Humeri*, as in *Men.*

c. The Tendinous Extremity of a fleshy Production of this *Muscle*, which is implanted on the Internal Protuberance of the *Os Humeri* of this *Animal*; as 'tis also in *Apes* and *Monkeys.*

The ninth Figure

Reprefents the *Urinary Parts* and *Organs* of *Generation.*

A. **T**HE left *Kidney* entire.

a. a. The *Membrana Adipofa*, partly freed from the *Kidney*, and turned back.

B. The *Right Kidney* opened, to shew its *Glandulous* Substance, and *Urinary Tubes*, and the *Pelvis.*

b The *Tubuli Urinarij* which arife from the *Glandulous* Substance, and like Lines drawn from a *Circumference* to a *Center*, pass to the *Fimbria* or Edge *c c*, in *Man* to the several *Papillæ*, where their Orifices open and empty themselves into the *Pelvis.*

c. c. The said *Fimbria*, of a Semicircular Figure, where the Extreams of the *Urinary Tubes* difcharge the *Urine* into the *Pelvis*, or rather *Funnel* of the *Kidneys.*

d. The *Pelvis* or *Infundibulum* : For being large here in the *Kidney*, and running into a long flender Stem in the *Ureter*, it more properly reprefents a *Funnel*, and ferves for the Conveying the *Urine* thence into the *Bladder.*

C. C. The *Glandulæ Renales.*

D. D. The Defcending Trunk of the *Arteria Magna* or *Aorta*, below the *Diaphragm.*

d. The *Cœliac Arterie.*

δ. The *Arteria Mefenterica fuperior.*

g The *Arteria Mefenterica inferior.*

E. The Defcending Trunk of the *Vena Cava.*

F. F. The *Emulgent Arteries.*

f. f. The *Emulgent Veins.*

G. G. The

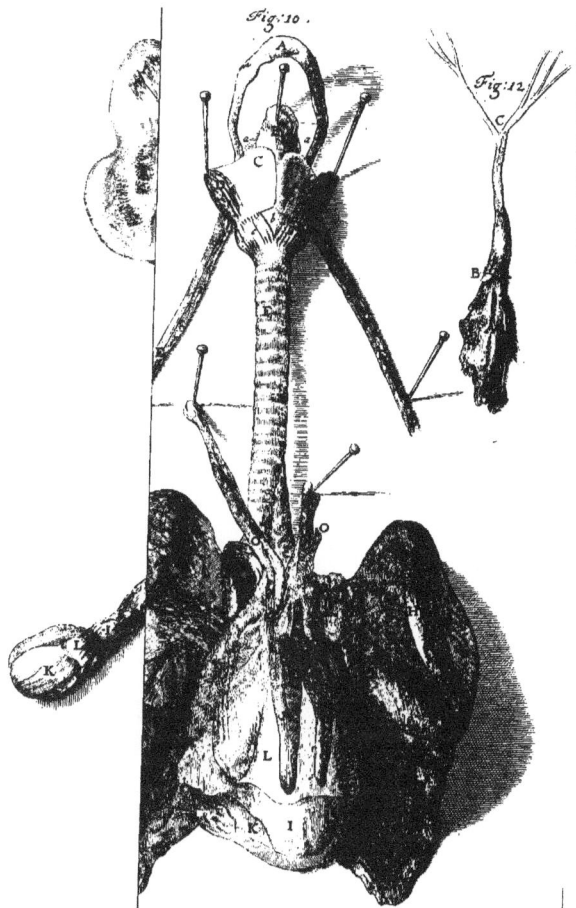

Fig: 10.

Fig: 12.

M. Vander Gucht Scul:

Fig: 9. Fig: 11. Fig: 10. Fig: 12.

M. Sandenynkt fecit.

G.G. The *Ureters.*

H. The *Bladder* of *Urine.*

h. h. The *Spermatick Veins* which difcharge themfelves into the *Vena Cava*, and the left *Emulgent*, as in *Man.*

j. The *Spermatick Arteries*, as they arife from the fore-part of the Trunk of the *Aorta.*

J. J. The *Vafa Præparantia Pampini-formia,feu Corpora Pyramidalia.*

K.K. The *Teftes* or *Stones*, which appear here flaccid, having been kept fome time, before the *figure* was taken.

L. L. The *Epididymis*, making feveral Convolutions on the body of the *Teftes.*

M. Part of the *Cremafter Mufcle.*

N. N. The *Vafa Deferentia.*

O. O. The *Veficulæ Seminales.*

P. The *Proftates* or *Corpus Glandofum.*

Q. The *Mufculus Erector Penis* of the right fide.

R. The upper part or *Dorfum Penis.*

S. The *Corpora Cavernofa Penis*, cut tranfverfe.

T. The *Urethra.*

k. k. The main Trunk of the *Iliac Arterie* and *Vein.*

l. l. The *Umbilical Arteries.*

m. m. The *Arterie* that goes to the *Penis.*

n. n. The *Arterie* that goes to the *Bladder* of Urine.

o. The internal *Iliac Vein* and *Arterie.*

p. The external *Iliac Vein* and *Arterie.*

q. The *Vena Pudenda feu Penis.*

r. r. The *Nerves* of the *Penis.*

f. f. The *Arteries* of the *Penis.*

The tenth Figure

Demonftrates the Parts of the *Thorax* with the *Arteria Afpera* and *Larynx.*

A. THE fore-part of the *Os Hyoides.*

a. a. Its two ends,that are connected to the two *Superior* long *Proceffes* of the *Cartilago Scutiformis.*

B. The *Epiglottis.*

C. The *Cartilago Scutiformis.*

b. The Prominent part of the *Annulary Cartilage.*

D. D. The *Mufculi Hyothyroidei.*

E. E. The *Mufculi Sternothyroidei.*

c. c. The *Mufculi Cricothyroidei.*

F. The *Arteria Afpera*, or *Wind-pipe.*

G. It's divifion,where it paffes to the right and left *Lobes* of the *Lungs.*

H. H. The *Lungs.*

P   *J.The*

J. The *Cone* of the *Heart.*

K. The right *Ventricle* of the *Heart* here opened, ſo that part of the *Polypus* contained there, came in view.

L. Part of the *Pericardium*, on the *Baſis* or upper part of the *Heart.*

M. M. The *Thymus*, lying on the *Pericardium.*

N. The *Mediaſtinum* freed from the *Sternum*, and turn'd to the right ſide.

O. O. The two *Subclavian Arteries.*

P. The *Carotid Arteries.*

The eleventh Figure

Shews the *Polypus* or Coagulated Blood found in the *left Ventricle* of the *Heart.*

A. THAT part contained in the *Ventricle.*

    B. Three Impreſſions, formed by the *Semilunary Valves.*

C. That part, that lay in the *Aorta.*

D. That part that paſſed into the deſcending Trunk of the *Aorta.*

E. Thoſe *Ramuli* of it that lay in the aſcendent Branches of the *Aorta.*

The twelfth Figure

The *Polypus* found in the right *Ventricle* of the *Heart.*

A. THAT part contained in the *Ventricle.*

    B. The Impreſſions made by the *Valvulæ ſigmoides.*

C. The Branches leading to the right and left *Lobes* of the *Lungs.*

The thirteenth Figure

Repreſents the *Baſis* of the *Brain* with the *Medulla Oblongata*, and the *Nerves* and *Arteries* cut off.

A. A. THE two *anterior* or fore *Lobes* of the *Brain.*

    B. B. The two *poſterior* or hinder *Lobes* of the *Brain.*

a.a. Two depreſſions in the fore *Lobes* cauſed by the riſing of the *Frontal* bone, that compoſes the upper part of the *Orbit* of the *Eyes*; which in this *Animal*, and in *Monkeys*, is more eminent than in *Man.*

b. b. The diviſion of the right and left *Hemiſphere* of the *Brain*, where the *Falx* is placed. This fore-part of the *Brain* in this *Animal* appeared ſomewhat flatter than in *Man.*

C. C. The *Cerebellum.*

D. The *Principium Medullæ Spinalis*, or that part of the *Caudex Medullaris,*

Fig. 16

M. Vander Gucht Sculp:

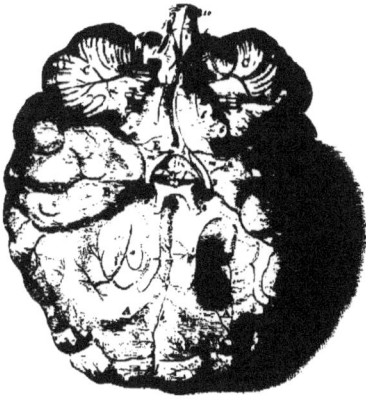

Fig. 13

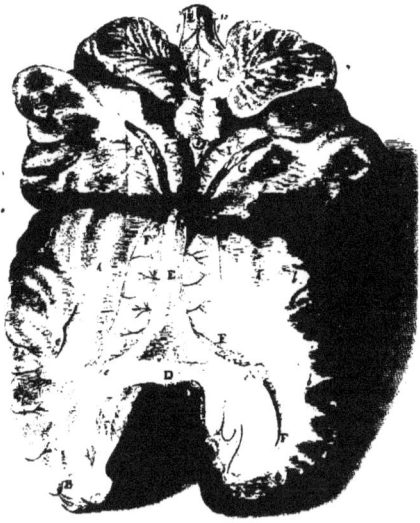

Fig. 14

Fig. 16

Fig. 17

Fig. 15

M Vander Gucht Sculp:

dullaris, where the *Corpora Pyramidalia* and *Olivaria* are placed, as in an Humane *Brain.*

E. E. The *Protuberantia Annularis,* or *Pons Verolij.*

e. e. The *Carotid Arteries.*

f. f. The *Vertebral Arteries.*

g. The *Cervical Arterie.*

h. The *Communicant* Branches between the *Cervical* and *Carotid Arteries.*

j. A small *Arterie* descending down the *Spinal Marrow.*

k. The *Infundibulum.*

l. l. The *Glandulæ duæ albæ pone Infundibulum,* , or rather two *Medullary. Protuberances* there.

m. m. Parts of the *Crura Medullæ Oblongatæ* before they unite under the *Pons Verolij,* or *Annular Protuberance.*

1. The *Olfactory,* or first pair of Nerves.

2. The *Optick,* or second pair of Nerves.

3. The *Nervi Oculorum motorij,* or third pair of Nerves.

4. The *Pathetick,* or fourth pair of Nerves.

5. The fifth pair of Nerves.

6. The sixth pair of Nerves.

7. The *Auditory,* or seventh pair of Nerves.

8. The *Par Vagum,* or eighth pair of Nerves.

9. The ninth pair of Nerves.

10. The tenth pair of Nerves, which may be reckoned rather the first pair of the *Neck;*

* * The *Nervus accessorius,* that goes to the eighth pair, or *Par Vagum.*

The fourteenth Figure.

Represents the inward Parts of the *Brain* , as divided by an *Horizontal Section*; where the *Basis* of the *Brain* is reflected upwards.

A. A. **P**Arts of the hinder *Lobes* of the *Brain.*

   B. B. The upper part of the *Brain* next it's *Hemispheres,* divided from the lower.

   C. C. The lower part next the *Basis,* reflected or turned up.

   a. a. The *Cortical* or *Cinericious* part of the *Brain,* which is *Glandulous.*

   b. b. The *Medullary* part, that runs up between the *Cortical,* and is *Nervous.*

   D. The *Corpus Transversale.*

   E. The *Fornix.*

   e. e. The *Crura Fornicis.*

   f. The two *Roots* of the *Fornix.*

· P 2   .F. F. The.

· *F. F.* The two firſt *Ventricles* of the *Brain.*

G.G. Parts of the *Corpora Striata,* entire.

g. g. The *Striæ* as they appear in this Section in the *Corpora Striata* in the lower part of the *Brain.*

h. h. The ſame *Striæ,* in the upper part of the *Brain.*

H. H. The *Plexus Choroides.*

J. J. The *Thalami Nervorum Opticorum.*

j. The *Plexus Choroides* continued.

K. K. The *Cerebellum* divided perpendicularly, to ſhew the ramifications of the *Medullary* part in the *Cortical.*

k. The *Foramen anterius* that leads to the Cavity under the *Protuberantiæ orbiculares.*

l. The *Glandula Pinealis.*

m. The *Nates.*

n. The *Teſtes.*

o. The *Commiſſure* of the *Medullary Proceſſes* of the *Cerebellum* and *Teſtes.*

p. The fourth *Ventricle* opened.

q. q. The *Acceſſory Nerves.*

10. The tenth pair of Nerves.

r. The *Foramen poſterius* or *inferius,* that leads to the Cavity under the *Protuberantiæ orbiculares.*

s. The *Rima* or *Foramen,* that leads to the *Infundibulum.*

The fiſteenth Figure

Is a Copy of the *Figure* that *Nicholaus Tulpius* gives of the *Orang-Outang* that was brought to *Holland* from *Angola.*

The fixteenth Figure

Repreſents the *Figure* that *Jacob. Bontius* gives of the *Orang-Outang* in *Piſo.*

The ſeventeenth Figure.

Is taken out of *Gefner,* which he tells us, he met with in a *German* Book, wrote about the *Holy Land.*

F I N I S.

A
PHILOLOGICAL
ESSAY

Concerning the

PYGMIES,

THE

CYNOCEPHALI,

THE

SATYRS and SPHINGES

OF THE

ANCIENTS,

Wherein it will appear that they were all either A P E S or M O N K E Y S; and not M E N, as formerly pretended.

By *Edward Tyson* M. D.

A

Philological Eſſay

Concerning the

P Y G M I E S

OF THE

A N C I E N T S.

HAVING had the Opportunity of Diſſecting this remarkable Creature, which not only in the *outward ſhape* of the
Body, but likewiſe in the ſtructure of many of the Inward
Parts, ſo nearly reſembles a *Man*, as plainly appears by the
Anatomy I have here given of it, it ſuggeſted the Thought
to me, whether this ſort of *Animal*, might not give the Foundation to
the Stories of the *Pygmies?* and afford an occaſion not only to the *Poets*,
but *Hiſtorians* too, of inventing the many *Fables* and wonderful and merry Relations, that are tranſmitted down to us concerning them? I muſt
confeſs, I could never before entertain any other Opinion about them,
but that the whole was a *Fiction:* and as the firſt Account we have of
them, was from a *Poet*, ſo that they were only a Creature of the Brain,
produced by a warm and wanton Imagination, and that they never had
any Exiſtence or Habitation elſewhere.

In this Opinion I was the more confirmed, becauſe the moſt diligent
Enquiries of late into all the Parts of the inhabited World, could never
diſcover any ſuch *Puny* diminutive *Race* of *Mankind*. That they ſhould
be totally deſtroyed by the *Cranes*, their Enemies, and not a Straggler
here and there left remaining, was a Fate, that even thoſe *Animals* that
are conſtantly preyed upon by others, never undergo. Nothing therefore appeared to me more Fabulous and Romantick, than their *Hiſtory*,
and the Relations about them, that *Antiquity* has delivered to us. And
not

not only *Strabo* of old, but our greatest Men of Learning of late, have wholly exploded them, as a meer *figment*; invented only to amuse, and divert the Reader with the Comical Narration of their Atchievements, believing that there were never any such Creatures in Nature.

This Opinion had so fully obtained with me, that I never thought it worth the Enquiry, how they came to invent such Extravagant Stories: Nor should I now, but upon the Occasion of Dissecting this *Animal:* For observing that 'tis call'd even to this day in the *Indian* or *Malabar* Language, *Orang-Outang*, *i.e.* a *Man* of the *Woods*, or *Wild-men*; and being brought from *Africa*, that part of the World, where the *Pygmies* are said to inhabit; and it's present *Stature* likewise tallying so well with that of the *Pygmies* of the Ancients; these Considerations put me upon the search, to inform my self farther about them, and to examine, whether I could meet with any thing that might illustrate their *History*. For I thought it strange, that if the whole was but a meer Fiction, that so many succeeding Generations should be so fond of preserving a *Story*, that had no Foundation at all in Nature; and that the *Ancients* should trouble themselves so much about them. If therefore I can make out in this *Essay*, that there were such *Animals* as *Pygmies*; and that they were not a *Race* of *Men*, but *Apes*; and can discover the *Authors*, who have forged all, or most of the idle Stories concerning them; and shew, how the Cheat in after Ages has been carried on, by embalming the Bodies of *Apes*, then exposing them for the *Men* of the Country, from whence they brought them: if I can do this, I shall think my time not wholly lost, nor the trouble altogether useless, that I have had in this Enquiry.

My Design is not to justifie all the Relations that have been given of this *Animal*, even by Authors of reputed Credit; but, as far as I can, to distinguish Truth from Fable; and herein, if what I assert amounts to a Probability, 'tis all I pretend to. I shall accordingly endeavour to make it appear, that not only the *Pygmies* of the Ancients, but also the *Cynocephali*, and *Satyrs* and *Sphinges* were only *Apes* or *Monkeys*, not *Men*, as they have been represented. But the Story of the *Pygmies* being the greatest Imposture, I shall chiefly concern my self about them, and shall be more concise on the others, since they will not need so strict an Examination.

We will begin with the Poet *Homer*, who is generally owned as the first Inventor of the Fable of the *Pygmies*, if it be a Fable, and not a true Story, as I believe will appear in the Account I shall give of them. Now *Homer* only mentions them in a *Simile*, wherein he compares the Shouts that the *Trojans* made, when they were going to joyn Battle with the *Græcians*, to the great Noise of the *Cranes*, going to fight the *Pygmies:* he saith (*a*),

(*a*) *Homer*. Iliad. lib. 3. ver. 4.

"At

Ἅι τ᾽ ἐπεὶ ἒν χειμῶνα φύγον, ἠ ἀθέσφατον ὄμβρον
Κλαγγῇ ταί γε πέτονται ἐπ᾽ ὠκεανοῖο ῥοάων
Ἀνδράσι πυγμαίοισι φόνον ἠ κῆρα φέρεσαι. i. e.

Quæ ſimul ac fugere Imbres, Hyememque Nivalem
Cum magno Oceani clangore ferantur ad undas
Pygmæis pugnamque Viris, cædeſque ferentes.

Or as *Helius Eobanus Heſſus* paraphraſes the whole (*b*).

Poſtquam ſub Ducibus digeſta per agmina ſtabant
Quæque ſuis, Equitum turmæ, Peditumque Cohortes,
Obvia torquentes Danais veſtigia Troes
Ibant, ſublato Campum claniore replentes :
Non ſecus ac cuneata Gruum ſublime volantum
Agmina, dum fugiunt Imbres, ac frigora Brunæ,
Per Cœlum matutino clangore feruntur,
Oceanumque petunt, mortem exitiumque cruentum
Irrita Pigmæis moturis arma ferentes.

By ἀνδράσι πυγμαίοισι therefore, which is the Paſſage upon which they have grounded all their fabulous Relations of the *Pygmies*, why may not *Homer* mean only *Pygmies* or *Apes* like *Men*. Such an Expreſſion is very allowable in a *Poet*, and is elegant and ſignificant, eſpecially ſince there is ſo good a Foundation in Nature for him to uſe it , as we have already ſeen, in the *Anatomy* of the *Orang-Outang.* Nor is a *Poet* tied to that ſtrictneſs of Expreſſion, as an *Hiſtorian* or *Philoſopher* ; he has the liberty of pleaſing the Reader's Phancy, by Pictures and Repreſentations of his own. If there be a becoming likeneſs, 'tis all that he is accountable for. I might therefore here make the ſame *Apology* for him, as *Strabo* (*c*) do's on another account for his *Geography,* ἐ γὸ κατ᾽ ἄγνοιαν τῷ τοπικῶν λέγεται, ἀλλ᾽ ἡδονῆς ἠ τέρψεως χάριν, That he ſaid it, not thro' Ignorance, but to pleaſe and delight : Or, as in another place he expreſſes himſelf (*d*), ἐ γὸ κατ᾽ ἄγνοιαν ὃ ἱςορίας ἀπολιπτέον γίνεθαι τῦτο, ἀλλὰ τερψωδίας χάριν. *Homer* did not make this ſlip thro' Ignorance of the true *Hiſtory*, but for the Beauty of his *Poem.* So that tho' he calls them *Men Pygmies*, yet he may mean no more by it , than that they were like *Men.* As to his Purpoſe, 'twill ſerve altogether as well, whether this bloody Battle be fought between the *Cranes* and *Pygmæan Men*, or the *Cranes* and *Apes*, which from their Stature he calls *Pygmies*, and from their ſhape *Men* ; provided that when the *Cranes* go to engage, they make a mighty terrible noiſe, and clang enough to fright theſe little *Wights* their mortal Enemies. To have called them only *Apes*, had been

---

(*b*) *Homeri Ilias Latino Carmine reddita ab Helio Eobano Heſſo.*   (*c*) *Strab. Geograph.* lib.1.p.m.25.
(*d*) *Strabo ibid.* p. m. 30.

B   ſtat

flat and low, and leſſened the grandieur of the Battle.   But this *Periphra-*
ſis of them, ἄνδρες τυγμαῖοι, raiſes the Reader's Phancy, and ſurpriſes
him, and is more becoming the Language of an *Heroic Poem.*

But how came the *Cranes* and *Pygmies* to fall out ? What may be the
Cauſe of this Mortal Feud, and conſtant War between them ? For *Brutes,*
like *Men,* don't war upon one another, to raiſe and encreaſe their Glo-
ry, or to enlarge their Empire.   Unleſs I can acquit my ſelf herein, and
aſſign ſome probable Cauſe hereof, I may incur the ſame Cenſure as *Strabo*
(*e*) paſſed on ſeveral of the *Indian Hiſtorians,* ἐπεχαλινσαν ᵕ ἰς τὼ Ὁμηει-
κὼ τῇ Πυγμαίων γεανομαχίαν τειαθάμεις; εἴπονῖες, for reviewing the
Homerical Fight of the *Cranes* and *Pygmies,* which he looks upon only
as a fiction of the *Poet.*   But this had been very unbecoming *Homer* to
take a *Simile* (which is deſigned for illuſtration) from what had no
Foundation in Nature.   His *Betrachomyomachia,* 'tis true, was a meer
Invention, and never otherwiſe eſteemed : But his *Geranomachia* hath all
the likelyhood of a true Story.   And therefore I ſhall enquire now what
may be the juſt Occaſion of this Quarrel.

Athenæus (*f*) out of *Philochorus,*and ſo likewiſe *Ælian* (*g*) , tell us a
Story, That in the Nation of the *Pygmies* the Male-line failing, one
Gerana was their Queen ; a Woman of an admired Beauty , and whom
the Citizens worſhipped as a *Goddeſs* ; but ſhe became ſo vain and proud,
as to prefer her own, before the Beauty of all the other *Goddeſſes,* at
which they grew enraged ; and to puniſh her for her Inſolence, *Athe-*
næus tells us that 'twas *Diana,*but *Ælian* ſaith 'twas *Juno* that transform-
ed her into a *Crane,* and made her an Enemy to the *Pygmies* that worſhipped her before.   But ſince they are not agreed which *Goddeſs* 'twas, I
ſhall let this paſs.

Pomponius Mela will have it, and I think ſome others, that theſe
cruel Engagements uſe to happen, upon the *Cranes* coming to devour the
Corn the *Pygmies* had ſowed ; and that at laſt they became ſo victori-
ous,as not only to deſtroy their Corn,but them alſo : For he tells us:(†),
Fuere interiùs Pygmæi , minutum genus , & quod pro ſatis frugibus contra
Grues dimicando, defecit.   This may ſeem a reaſonable Cauſe of a Quarrel ; but it not being certain that the *Pygmies* uſed to ſow *Corn,* I will
not inſiſt on this neither.   ·

Now what ſeems moſt likely to me, is the account that *Pliny* out of
Megaſthenes, and *Strabo* from *Oneſicritus* give us ; and , provided I be
not obliged to believe or juſtifie *all* that they ſay, I could reſt ſatisfied in
great part of their Relation : For *Pliny* (*h*) tells us,*Veris tempore univerſo*

(e) *Strabo Geograph.* lib. 2. p. m. 48.   (f) *Athenæi Deipnoſoph.* lib. 9. p. m. 393.   (g) *Ælian.*
Hiſt. Animal. lib. 15. cap. 29.   (†) *Pomp. Mela de ſitu Orbis,* lib. 3. cap. 8.   (h) *Plinij Hiſt. Nat.*
lib. 7. cap. 2. p. m. 13

agmine

agmine ad mare defcendere, & *Ova, Pullofque carum Alitum confumere:* That in the Spring-time the whole drove of the *Pygmies* go down to the Sea fide, to devour the *Cranes* Eggs and their young Ones. So likewife *Oneficritus* (i), Πρὸς ὂ τὰς τριαπιθάμυς πόλεμον ἐι) ταῖς Γεράνοις (ὃν ἢ Ὅμηρον δηλοῦν) ἢ τοῖς Πέρδιξιν, ἃς χνομμγέθεις ἐι) τότες δ' ἀπλέγειν αὐτῷ τὰ ὠὰ, ἢ φθείρειν· ἐκεῖ γὸ ὠοτικεῖν τὰς Γεράνυς· δίπερ μηδαμῆ μηδ᾽ ὠὰ ἐυρίσκεθαι Γεράνων, μηⁿ ἂν νεότιτα· i. e. *That there is a fight between the* Pygmies *and the* Cranes *(as* Homer *relates) and the* Partridges, *which are as big as* Geefe *; for thefe* Pygmies *gather up their Eggs, and deftroy them; the* Cranes *laying their Eggs there; and neither their Eggs, nor their Nefts, being to be found any where elfe.* 'Tis plain therefore from them, that the Quarrel is not out of any *Antipathy* the *Pygmies* have to the *Cranes,* but out of love to their own Bellies. But the *Cranes* finding their Nefts to be robb'd, and their young Ones prey'd on by thefe Invaders, no wonder that they fhould fo fharply engage them; and the leaft they could do, was to fight to the utmoft fo mortal an Enemy. Hence, no doubt, many a bloody Battle happens, with various fuccefs to the Combatants; fome-times with great flaughter of the *long-necked Squadron;* fometimes with great effufion of *Pygmæan* blood. And this may well enough, in a *Poet*'s phancy, be magnified, and reprefented as a dreadful War; and no doubt of it, were one a *Spectator* of it, 'twould be diverting enough.

─────*Si videas hoc*
Gentibus in noftris, rifu quatiere: fed illic,
Quanquam eadem affiduè fpectantur Prælia, ridet
Nemo, ubi tota cohors pede non eft altior uno (k).

This Account therefore of thefe Campaigns renewed every year on this Provocation between the *Cranes* and the *Pygmies,* contains nothing but what a cautious Man may believe; and *Homer*'s *Simile* in likening the great fhouts of the *Trojans* to the Noife of the *Cranes,* and the Silence of the *Greeks* to that of the *Pygmies,* is very admirable and delightful. For *A-riftotle* (l) tells us, That the *Cranes,* to avoid the hardfhips of the Win-ter, take a Flight out of *Scythia* to the *Lakes* about the *Nile,* where the *Pygmies* live, and where 'tis very likely the *Cranes* may lay their Eggs and breed, before they return. But thefe rude *Pygmies* making too bold with them, what could the *Cranes* do lefs for preferving their Off-fpring than fight them; or at leaft by their mighty Noife, make a fhew as if they would. This is but what we may obferve in all other Birds. And thus far I think our *Geranomachia* or *Pygmæomachia* looks like a true Story; and there is nothing in *Homer* about it, but what is credible. He only expreffes himfelf, as a *Poet* fhould do; and if Readers will miftake his meaning, 'tis not his fault.

────────────

(i) *Strab. Geograph. lib.* 15. pag. 489.   (k) *Juvenal. Satyr.* 13. verf. 170.   (l) *Ariftotle. Hift.* *Animal.* lib. 8. cap. 15. Edit. Scalig.

'Tis

'Tis not therefore the *Poet* that is to be blamed, tho' they would fa-
ther it all on him ; but the fabulous *Historians* in after Ages, who have
so odly drest up this Story by their fantastical Inventions , that there is
no knowing the truth, till one hath pull'd off those Masks and Visages,
wherewith they have disguised it.   For tho' I can believe *Homer*, that
there is a fight between the *Cranes* and *Pygmies*, yet I think I am no ways
obliged to imagine, that when the *Pygmies* go to these Campaigns to
fight the *Cranes*, that they ride upon *Partridges*, as *Athenæus* from *Ba-
silis* an *Indian Historian* tells us ; for, saith he (*m*) , Βάσιλις ἢ ἐν τῷ δδυ-
τέρῳ τῷ Ἰνδικῶν, οἱ μικροὶ, φησὶν, ἄνδρες; οἱ ταῖς Γεράνοις διαπολεμῖσντες Πέρδιξιν
ὀχημαλι χρῶνlαι.   For presently afterwards he tells us from *Menecles* ,
that the *Pygmies* not only fight the *Cranes*, but the *Partridges* too, Μενέ-
κλῆς ἢ ἐν πρώτῃ τῆς συναγωγῆς οἱ πυῖμαῖοι, φησὶ, ταῖς πέρδιξι, ἢ ταῖς Γεράνοις
πολέμισn.   This I could more readily agree to, because *Onesicritus*, as I
have quoted him already confirms it ; and gives us the same reason for
this, as for fighting the *Cranes*, because they rob their Nests.   But whether these *Partridges* are as big as *Geese*, I leave as a *Quære*.

Megasthenes methinks in *Pliny* mounts the *Pygmies* for this Expedition
much better, for he sets them not on a *Pegasus* or *Partridges* , but on
Rams and *Goats : Fama est* (saith *Pliny* (*n*)) *insidentes Arietum Capra-
rumque dorsis, armatis sagittis, veris tempore universo agmine ad mare de-
scendere.*   And *Onesicritus* in *Strabo* tells us, That a *Crane* has been often
observed to fly from those parts with a brass Sword fixt in him , πλεισ᾽άκις
δ᾽ ἐπῖπῖειν γέρανον χανλω ἔχυσαν ἀκίδα ἀπὸ τῷ ὀκεῖθεν πληῖμάτων (*o*).
But whether the *Pygmies* do wear Swords, may be doubted.   'Tis true,
Ctesias tells us (*p*), That the *King* of India every fifth year sends fifty
Thousand Swords, besides abundance of other Weapons , to the Nation
of the *Cynocephali*, (a sort of *Monkeys*, as I shall shew) that live in those
Countreys, but higher up in the Mountains : But he makes no mention
of any such Presents to the poor *Pygmies* ; tho' he assures us, that no less
than three Thousand of these *Pygmies* are the *King's* constant Guards:
But withal tells us, that they are excellent *Archers*, and so perhaps by
dispatching their Enemies at a distance, they may have no need of such
Weapons to lye dangling by their sides.   I may therefore be mistaken in
rendering ἀκίδα a Sword ; it may be any other sharp pointed Instrument
or Weapon, and upon second Thoughts, shall suppose it a sort of Ar-
row these cunning *Archers* use in these Engagements.

These, and a hundred such ridiculous *Fables,* have the *Historians* invented of the *Pygmies,* that I can't but be of *Strabo's* mind (*q*), Ῥᾴδιον δ᾽
ἄν τις Ἡσιόδῳ, ἢ Ὁμήρῳ πιςδύσειεν ἡρωολογῦσι , ἢ ταῖς τραγικοῖς ποιηλαῖς, ἢ
Κῖησία τὶ ἢ Ἡροδότῳ, ἢ Ἑλλανίκῳ, ἢ ἄλλοις τοιέτοις' i. e. *That one may soon-
er believe* Hesiod, *and* Homer, *and the* Tragick Poets *speaking of their*

(m) *Athenæi Diepnosoph.* lib. p. 9 . m. 350.   (n) *Plinij Nat. Hist.* lib.7.cap.2. p. 13.   (o) *Strabo*
Geograph. lib. 15. p. 489.   (p) *Vide Photij Biblioth.*   (q) *Strabo Geograph.* lib. 11. p. m. 350.

Hero's,

Hero's, *than* Ctefias *and* Herodotus *and* Hellanicus, *and fuch like.* So ill an Opinion had *Strabo* of the *Indian Hiftorians* in general, that he cenfures them *all* as fabulous (r); Ἅπαντες μὲν τοίνυν ὡ περὶ τ̃ Ἰνδικῆς γεγ̓αϕότες ὡς ὃπὶ τὸ πολὺ ψδδολόγοι γεγόνασι, καθ᾿ ὑπερβολὼ ᾗ Δαίμαχℴ· τὰ ᾗ δ᾿περα λέγει Μεγαϑίνης, Ὀνησικριτός τι ᾗ Νέαρχℴ, ᾗ ἄλλοι τοιῆτοι· i. e. *All who have wrote of* India, *for the moft part, are fabulous , but in the higheft degree* Daimachus ; *then* Megafthenes, Onefícritus, *and* Nearchus, *and fuch like.* And as if it had been their greateft Ambition to excel herein, *Strabo* (s) brings in *Theopompus,* as bragging, Ὅτι ᾗ μύθℴς ἐν ταῖς Ἰσορίαις ἐρεῖ κρεῖπον, ἢ ὡς Ἡρόδοτℴ, ᾗ Κτησίας, ᾗ Ἑλλάνικℴ, ᾗ οἱ τὰ Ἰνδικὰ συγρά̓ψαντες· *That he could foift in Fables into Hiftory, better than* Herodotus *and* Ctefias *and* Hellanicus, *and all that have wrote of* India. The *Satyrift* therefore had reafon to fay,

——*Et quicquid Græcia mendax*
Audet in Hiftoria (t).

Ariftotle (u), 'tis true, tells us, Ὅλως ᾗ τὰ μὲν ἄγρια ἀγριώτερα ἐν τῇ Ἀσία, ἀνδρειότερα ᾗ πάντα τὰ ἐν τῇ Εὐρώπη, πολυμορϕότατα ᾗ τὰ ἐν τῇ Λιβύη· ᾗ λέγε͂αι δή τις παρριμία, ὅτι ἀεὶ ϕέρει τι Λιβύη καινόν· i. e. *That ge-nerally the Beafts are wilder in* Afia, *ftronger in* Europe, *and of greater va-riety of fhapes in* Africa ; *for as the Proverb faith ,* Africa *always produces fomething new.* Pliny (w) indeed afcribes it to the Heat of the *Climate, Animalium, Hominumque effigies monftriferas, circa extremitates ejus gigni, minimè mirum, artifici ad formanda Corpora, effigiefque cælandas mobilitate igneâ.* But *Nature* never formed a whole *Species* of *Monfters* ; and 'tis not the *heat* of the Country, but the warm and fertile Imagination of thefe *Hiftorians,* that has been more productive of them, than *Africa* it felf ; as will farther appear by what I fhall produce out of them, and particularly from the Relation that *Ctefias* makes of the *Pygmies.*

I am the more willing to inftance in *Ctefias,* becaufe he tells his Story roundly ; he no ways minces it ; his Invention is ftrong and fruitful ; and that you may not in the leaft miftruft him, he pawns his word, that all that he writes, is certainly true : And fo fuccefsful he has been, how Romantick foever his Stories may appear, that they have been handed down to us by a great many other Authors, and of Note too ; tho' fome at the fame time have look'd upon them as meer Fables. So that for the prefent, till I am better informed, and I am not over curious in it, I fhall make *Ctefias,* and the other *Indian Hiftorians,* the *Inventors* of the extravagant Relations we at prefent have of the *Pygmies,* and not old *Homer.* He calls them, 'tis true, from fomething of Refemblance of their fhape, ἄνδρες : But thefe *Hiftorians* make them to fpeak the *Indian Language* ; to ufe the fame *Laws* ; and to be fo confiderable a Nation,

(r) *Strabo ibid.* lib. 2. p. m. 48.   (s) *Strabo ibid.* lib. 1. p. m. 29.   (t) *Juvenal. Satyr.* X. verf. 174.   (u) *Ariftotle Hift. Animal.* lib. 8. cap. 28.   (w) *Plin. Nat. Hift.* lib. 6. cap. 30. p. m. 741.

and

and so valiant, as that the *King* of *India* makes choice of them for his *Corps de Guards*; which utterly spoils *Homer's Simile*, in making them so little, as only to fight *Cranes*.

Ctesias's Account therefore of the *Pygmies*(as I find it in *Photius's Bibliotheca* (x)), and at the latter end of some Editions of *Herodotus*) is this :

Ὅτι ἐν μέσῃ τῇ Ἰνδικῇ ἄνθρωποι εἰσὶ μέλανες, ἢ καλοῦνται πυγμαῖοι, τοῖς ἄλλοις ὁμόγλωσσοι Ἰνδοῖς· μικροὶ δέ εἰσι λίαν· οἱ μακρότατοι αὐτῶν πήχεων δύο, οἱ δὲ πλεῖσοι, ἑνὸς ἡμίσεος πήχεως, κόμην δὲ ἔχειν μακροτάτην, μέχρι ἢ ἐπὶ τὰ γόνατα, ἢ ἔτι καλώτερον, ἢ πώγωνα μέγιστον πάντων ἀνθρώπων· ἐπειδὰν οὖν τὸν πώγωνα μέγα φύσωσιν, οὐκέτι ἀμφιέννυνται οὐδὲν ἱμάτιον : ἀλλὰ τὰς τρίχας, τὰς μὲν ἐκ τῆς κεφαλῆς, ὄπισθεν καθιεῖσαι πολὺ κάτω τῶν γονάτων· τὰς δὲ ἐκ τοῦ πώγωνος, ἔμπροσθεν μέχρι ποδῶν ἑλκομένας. Ἔπειτα περιπυκνωσάμενοι τὰς τρίχας περὶ ἅπαν τὸ σῶμα, ζώννυνται, χρώμενοι αὐταῖς ἀντὶ ἱματίων. αἰδοῖον δὲ μέγα ἔχουσιν, ὥστε ψαύειν τῶν σφυρῶν αὐτῶν, ἢ παχύ. αὐτοί τε σιμοί τε ἢ αἰσχροί. τὰ δὲ πρόβατα αὐτῶν, ὡς ἄρνες. ἢ αἱ βόες ἢ οἱ ὄνοι, σχεδὸν ὅσον κριοί. ἢ οἱ ἵπποι αὐτῶν ἢ οἱ ἡμίονοι, ἢ τὰ ἄλλα πάντα ζῶα, οὐδὲν μεῖζω κριῶν. ἕπονται δὲ τῷ βασιλεῖ τῶν Ἰνδῶν, τέταν τῶν πυγμαίων ἄνδρες τρισχίλιοι. σφόδρα γάρ εἰσι τοξόται· δικαιότατοι δέ εἰσι καὶ νόμοισι χρῶνται ὥσπερ καὶ οἱ Ἰνδοί. Λαγωὸς τε καὶ ἀλώπεκας θηρεύουσιν, οὐ τοῖς κυσὶν, ἀλλὰ κόραξι καὶ ἰκτῖσι καὶ κορώναις καὶ ἀετοῖς.

Narrat praeter ista, in media India homines reperiri nigros, qui Pygmaei appellantur. Eadem hos, qua Inda reliqui, lingua uti, sed valde esse parvos, ut maximi duorum cubitorum, & plerique unius duntaxat cubiti cum dimidio altitudinem non excedant. Comam alere longissimam, ad ipsa usque genua demissam, atque etiam infra, cum barba longiore, quàm apud ullos hominum. Quae quidem ubi illis promissior esse coeperit, nulla deinceps veste uti : sed capillos nullò infra genua à tergo demissos, barbámque praeter pectus ad pedes usque defluentem, per totum corpus in orbem constipare & cingere, atque ita pilos ipsis suos vestimenti loco esse. Veretrum illis esse crassum ac longum, quod ad ipsos quoque pedum malleolos pertingat. Pygmeos hosce simis esse naribus, & deformes. Ipsorum item oves agnorum nostrorum instar esse; boves & asinos, arietum ferè magnitudine, equos item multósque & caetera jumenta omnia nihilo esse nostris arietibus majora. Tria horum Pygmaeorum millia Indorum regem in suo comitatu habere, quòd sagittarii sint peritissimi. Summos esse justitiae cultores, iisdémque quibus Indi reliqui, legibus parere.

Venari quoque lepores vulpésque, non canibus, sed corvis, milvis, cornicibus, aquilis adhibitis.

' In the middle of *India* (saith *Ctesias*) there are black Men, they are ' call'd *Pygmies*, using the same Language , as the other *Indians* ; they

(x) *Photij Bibliothec. Cod. 72. p. m. 145.*

are

' are very little , the talleſt of them being but two Cubits, and moſt of
' them but a Cubit and a half high. They have very long hair, reaching
' down to their Knees and lower ; and a Beard larger than any Man's.
' After their Beards are grown long, they wear no Cloaths, but the Hair
' of their Head falls behind a great deal below their Hams ; and that of
' their Beards before comes down to their Feet: then laying their Hair
' thick all about their Body, they afterwards gird themſelves, making
' uſe of their Hair for Cloaths. They have a *Penis* ſo long, that it rea-
' ches to the Ancle, and the thickneſs is proportionable. They are flat
' noſed, and ill favoured. Their Sheep are like Lambs ; and their Oxen
' and Aſſes ſcarce as big as Rams ; and their Horſes and Mules, and all
' their other Cattle not bigger. Three thouſand Men of theſe *Pygmies*
' do attend the *King* of *India*. They are good *Archers* ; they are very
' juſt, and uſe the ſame *Laws* as the *Indians* do. They kill Hares and
' Foxes, not with Dogs, but with Ravens, Kites, Crows, and Eagles.

Well; if they are ſo good Sports-men, as to kill Hares and Foxes with
Ravens, Kites, Crows and Eagles, I can't ſee how I can bring off *Homer*,
for making them fight the *Cranes*-themſelves. Why did they not fly
their *Eagles* againſt them ? theſe would make greater Slaughter and Execution, without hazarding themſelves. The only Excuſe I have is, that
Homer's Pygmies were real *Apes* like *Men* ; but thoſe of *Cteſias* were neither *Men* nor *Pygmies* ; only a Creature begot in his own Brain, and to
be found no where elſe.

Cteſias was Phyſician to *Artaxerxes Mnemon* as *Diodorus Siculus* (*y*)
and *Strabo* (*z*) inform us. He was contemporary with *Xenophon*, a little
later than *Herodotus* ; and *Helvicus* in his *Chronology* places him three hundred eighty three years before *Chriſt :* He is an ancient Author, 'tis true,
and it may be upon that ſcore valued by ſome. We are beholden to him,
not only for his Improvements on the Story of the *Pygmies*, but for his
Remarks likewiſe on ſeveral other parts of *Natural Hiſtory* ; which for
the moſt part are all of the ſame ſtamp, very wonderful and incredible ;
as his *Mantichora*, his *Gryphins*, the *horrible Indian Worm*, a Fountain of
Liquid Gold, a Fountain of *Honey*, a Fountain whoſe Water will make
a Man confeſs all that ever he did, a Root he calls πάρηβον, that will at-
tract Lambs and Birds, as the Loadſtone does filings of Steel ; and a great
many other Wonders he tells us: all which are copied from him by
Ælian, *Pliny*, *Solinus*, *Mela*, *Philoſtratus* and others. And *Photius* concludes *Cteſias's* Account of *India* with this paſſage ; Ταῦτα γεάφων καὶ
μυϑολογῶν Κτησίας. Λίγει τ' ἀληϑίςαλα γεάφειν' ἐπάγων ὡς τὰ μὲν αὐτὸς ἰδὼν
γεάφει, τὰ ϑ' παρ' αὐτῶ μαϑὼν τῶ εἰδότων. πολλὰ ϑ' τότων καὶ ἄλλα ϑαυ-
μασιώτεεα παραλιπεῖν, διὰ τὸ μὴ δόξαι τοῖς μὴ ταῦτα ϑεασαμένοις ἄπιςα
συγγράφειν' i. e. *Theſe things* (ſaith he) Cteſias *writes and feigns, but he*

(y) *Diodor. Siculi Bibliothec.* lib. 2. p. m. 118.   (z) *Strabo Geograph.* lib. 14. p. 451.

him

himself says all he has wrote is very true. Adding, *that some things which he describes, he had seen himself ; and the others, he had learn'd from those that had seen them : That he had omitted a great many other things more wonderful, because he would not seem to those that have not seen them, to write incredibilities.* But notwithstanding all this, *Lucian* (*a*) will not believe a word he saith ; for he tells us that *Ctesias* has wrote of *India* , Ἅ μήτε αὐτὸς εἶδε, μήτε ἄλλε εἰπόντθ᾽ ἤκυσεν, *What he neither saw himself, nor ever heard from any Body else.* And *Aristotle* tells us plainly, he is not fit to be believed : Ἐν δὲ τῇ Ἰνδικῇ ὥς φησι Κτησίας, ἐκ ἂν ἀξιόπισθ᾽ (*b*). And the same Opinion *A. Gellius* (*c*) seems to have of him, as he had likewise of several other old *Greek Historians* which happened to fall into his hands at *Brundusium*, in his return from *Greece* into *Italy* ; he gives this Character of them and their performance : *Erant autem isti omnes libri Græci, miraculorum fabularumque pleni : res inauditæ, incredulæ, Scriptores veteres non parvæ authoritatis,* Aristeas Proconnesius, & Isagonus, & Nicæensis, & Ctesias, & Onesicritus, & Polystephanus, & Hegesias. Not that I think all that *Ctesias* has wrote is fabulous ; For tho' I cannot believe his *speaking Pygmies*, yet what he writes of the *Bird* he calls Βίτϊακ.θ᾽, that it would speak *Greek* and the *Indian Language*, no doubt is very true ; and as *H. Stephens* (*d*) observes in his Apology for *Ctesias*, such a Relation would seem very surprising to one, that had never seen nor heard of a *Parrot*.

But this Story of *Ctesias's speaking Pygmies*, seems to be confirm'd by the Account that *Nonnosus*, the Emperour *Justinian's* Ambassador into *Æthiopia*, gives of his Travels. I will transcribe the Passage, as I find it in *Photius* (*e*), and 'tis as follows :

Ὅτι ἀπὸ τῇ φερσαῖν πλέοντι τῷ Νον-νίσῳ, ὑπὶ τῇ ἐχάτλω τῷ νήσων καλλωλη-κότι ποῖεν δέ τι συνίδη, θαῦμα ἢ ἀκέσαι. ἀείτυκε γάρ τισι μορφλώ μὲν ἢ ἰσχε-αν ἔχων ἀνθρωπίνω, βραχυλάτοις δ᾽ τὸ μέγαθθ᾽, ἢ μέλαν τ᾽ χεςαν. ἀπὸ δ᾽ τεςγχαν δεδασυπλάντις διὰ παντὸς τῦ σώ καλθ᾽. Εἴπετὶ δ᾽ τοῖς αὐδεσιν ἢ γωνᾶικες παςαπλήσιαι ἢ παιδάεια ἐπι βραχύτεςα, τ᾽ῆ παρ᾽ αὐτοῖς ἀνδεῶν. γυμνοὶ δ᾽ ἦσαν ἅπαντες᾽ πλὴν δέματι τινὶ μικρᾷ τ᾽ αἰδῶ πεςικεκαλυπτεῖν, οἱ προ᾽εβήκλης ὁμοίως ἀνδεες τε καὶ γυναῖκες. ἄγειον δ᾽ ἐδὲν ἐπεδείκνυλο ἐδὲ ἀνήμεφον᾽ ἀλλὰ καὶ φωνὴ εἴχον μὲν ἀνθρωπίνω, ἄλλωϛεν δ᾽ πανἰάπασι τ᾽	*Naviganti à Pharsa Nonnoso, & ad extremam usque insularum delato, tale quid occurrit, vel ipso auditu admirandum.* Incidit enim in quosdam forma quidem & figura humana, sed brevissimos, & cutem nigros, totúmque pilosos corpus. *Sequebantur viros æquales fæminæ, & pueri adhuc breviores.* Nudi omnes agunt, pelle tantum brevi adultiores verenda tecti, viri pariter ac fæminæ: agreste nihil, neque efferum quid præ se ferentes. Quin & vox illis humana, sed omnibus, etiam accolis, prorsus ignota lingua, multoque amplius Nonnosi sociis. Vivunt marinis ostreis, & piscibus è

(a) *Lucian lib.* 1. *vera Histor.* p. m. 373.   (b) *Arist. Hist. Animal.* lib. 8. cap. 28.   (c) *A. Gellij Noctes Attic.* lib. 9. cap. 4.   (d) *Henr. Stephani de Ctesia Historico antiquissimo disquisitio, ad finem Herodoti.*   (e) *Photij Bibliothec.* cod. 3. p. m. 7.

                                                              διάλεκτον

διάλεκτον τοῖς τε πεςοίκοις ἅπασι, καὶ
πολλῷ πλέον τοῖς πεςὶ τ̂ Νοννοσον, Δέ-
ζων ἢ ἐκ θαλαττ̂ίων ὀςρείων, κ̀ ἰχθύων,
τ̀ ἀπὸ τ̂ θαλάσσης εἰς τ̂ νῆσον ἀποῤῥιπ-
τομ́νων· Θάρσος ἢ εἶχον ἐδὲν. ἀλλὰ κ̀
ὁρῶντες τὲς καθ᾽ ἡμᾶς ἀνθρώπες ὑπέπησαν, ὥσπες ἡμεῖς τὰ μείσω τ̂ ᾿ Θηρίων.

é mari ad insulam projectis. Auda-
ces minimè sunt, ut nostris conspectis
hominibus, quemadmodum nos visa
ingenti fera, metu perculsi fuerint.

‘ That _Nonnosus_ sailing from _Pharsa_, when he came to the farthermost
‘ of the Islands, a thing, very strange to be heard of, happened to him ;
‘ for he lighted on some _(Animals)_ in shape and appearance like _Men_,
‘ but little of stature, and of a black colour, and thick covered with
‘ hair all over their Bodies. The Women, who were of the same sta-
‘ ture, followed the Men : They were all naked, only the Elder of them,
‘ both Men and Women, covered their Privy Parts with a small Skin.
‘ They seemed not at all fierce or wild ; they had a Humane Voice, but
‘ their _Dialect_ was altogether unknown to every Body that lived about
‘ them ; much more to those that were with _Nonnosus._ They liv'd upon
‘ Sea Oysters, and Fish that were cast out of the Sea, upon the Island.
‘ They had no Courage ; for seeing our Men, they were frighted, as we
‘ are at the sight of the greatest wild Beast.

Φωνὴὼ εἶχον μὲν ἀνθρωπίνἠω I render here, _they had a Humane Voice,_ not
Speech : for had they spoke any Language, tho' their _Dialect_ might be
somewhat different, yet no doubt but some of the Neighbourhood would
have understood something of it, and not have been such utter Strangers
to it. Now 'twas observed of the _Orang-Outang,_ that it's _Voice_ was like
the _Humane,_ and it would make a Noise like a Child, but never was observed to speak, tho' it had the _Organs_ of _Speech_ exactly formed as they
are in _Man_ ; and no Account that ever has been given of this Animal
do's pretend that ever it did. I should rather agree to what _Pliny (f)_ mentions, _Quibusdam pro Sermone nutus motusque Membrorum est_ ; and that they
had no more a Speech, than _Ctesias_ his _Cynocephali_ which could only bark,
as the same _Pliny (g)_ remarks ; where he saith, _In multis autem Montibus_
Genus Hominum Capitibus Caninis , ferarum pellibus velari, pro voce latra-
tum edere, unguibus armatum venatu & Aucupio vesci, horum supra Centum
viginti Millia fuisse prodente se Ctesias scribit. But in _Photius_ I find , that
Ctesias's Cynocephali did speak the _Indian Language_ as well as the _Pygmies._
Those therefore in _Nonnosus_ since they did not speak the _Indian,_ I doubt,
spoke no _Language_ at all ; or at least, no more than other _Brutes_ do.

Ctesias I find is the only Author that ever understood what Language
'twas that the _Pygmies_ spake : For _Herodotus (h)_ owns that they use a
sort of Tongue like to no other, but screech like _Bats._ He saith, Οἱ Γα-
ραμαντες ἔτοι τὲς Τρωγλοδύτας Αἰθίοπας Θηρεύεσι τοῖσι τεθείπποισι. Οἱ γὸ

(f) _Plinij Nat. Hist._ lib. 6. cap. 30. p. m. 741.   (g) _Plinij Nat. Hist._ lib. 7. cap. 2. p. m. 11.
(h) _Herodot._ in _Melpomene._ pag. 283.
C
Τρω.

Τρωγλοδύται αἰθίοπες πόδας τάχιςοι ἀνθρώπων πάντων εἰσὶ, ᾗ ἡμεῖς περι λόγυς ἀποφερομένης ἀκήσαμεν. Σιτέονται ᾗ οἱ Τρωγλοδύται ὄφις, ἢ Σαύρης, ἢ τὰ τοιαῦτα ᾗ Ἑρπετῶν. Γλῶσσαν ᾗ ὀδεμιῇ ἄλλη παρομοίευ νενομίκασι, ἀλλὰ τέτριγασι καθάπερ αἱ νυκτερίδες· i. e. *These* Garamantes *hunt the* Troglodyte Æthiopians *in Chariots with four Horses. The* Troglodyte Æthiopi:ns *are the swiftest of foot of all Men that ever he heard of by any Report. The* Troglodytes *eat Serpents and Lizards, and such sort of Reptiles. They use a Language like to no other Tongue, but screech like Bats.*

Now that the *Pygmies* are *Troglodytes*, or do live in Caves, is plain from *Aristotle* (i), who faith, Τρωγλοδύται δ᾽ εἰσὶ τ̄ Βίον. And so *Philostratus* (k), Τὸς ᾗ πυΓμαίας οἰκεῖν μὲν ὑπογείυς. And methinks *Le Compte's* Relation concerning the *wild* or *savage Man* in *Borneo*, agrees so well with this, that I shall transcribe it : for he tells us , (l) *That* in Borneo *this* wild *or* savage Man *is indued with extraordinary strength ; and notwithstanding he walks but upon two Legs, yet he is so swift of foot, that they have much ado to outrun him. People of Quality course him , as we do Stags, here : and this sort of hunting is the King's usual divertisement.* And *Gassendus* in the Life of *Peiresky,* tells us they commonly hunt them too in *Angola* in *Africa*, as I have already mentioned. So that very likely *Herodotus*'s *Troglodyte Æthiopians* may be no other than our *Orang-Outang* or *wild Man.* And the rather, because I fancy their Language is much the same : for an *Ape* will chatter, and make a noise like a *Bat*, as his *Troglodytes* did : And they undergo to this day the same Fate of being hunted, as formerly the *Troglodytes* used to be by the *Garamantes.*

Whether those ἄνδρας μικρὸς μέϊε͜ιαν ἐλάσσονας ἀνδρῶν which the *Nasamones* met with (as *Herodotus* (m) relates) in their Travels to discover *Libya*, were the *Pygmies* ; I will not determine : It seems the *Nasamones* neither understood their Language, nor they that of the *Nasamones.* However, they were so kind to the *Nasamones* as to be their Guides along the Lakes, and afterwards brought them to a City, ἐν τῇ πάντας ἦ τοῖσι ἄγεσι τὸ μέγαθ Θ ἴσυς, χρῶμα ᾗ μέλανας, i. e. *in which all were of the same stature with the Guides, and black.* Now since they were all *little black Men*, and their Language could not be understood, I do suspect they may be a Colony of the *Pygmies :* And that they were no farther Guides to the *Nasamones*, than that being frighted at the sight of them, they ran home, and the *Nasamones* followed them.

I do not find therefore any good Authority, unless you will reckon *Ctesias* as such, that the *Pygmies* ever used a Language or Speech, any

----

(i) *Arist. Hist. Animal.* lib. 8. cap. 15. p. m. 913.   (k) *Philostrat. in vita Apollon. Tyanæi,* lib. 3. cap. 14. p. m. 152.   (l) *Lewis le Compte* Memoirs and Observations on *China,* p. m. 510.   (m) *Herodotus in Euterpe* seu lib. 2. p. m. 102.

more than other *Brutes* of the fame *Species* do among themfelves, and what we know nothing of, whatever *Democritus* and *Melampodes* in *Pliny* (*n*),or *Apollonius Tyanæus* in *Porphyry* (*o*) might formerly have done. Had the *Pygmies* ever fpoke any *Language* intelligible by Mankind, this might have furnifhed our *Hiftorians* with notable Subjeĉts for their *Novels*; and no doubt but we fhould have had plenty of them.

But *Albertus Magnus*, who was fo lucky as to guefs that the *Pygmies* were a fort of *Apes* ; that he fhould afterwards make thefe *Apes* to *fpeak*, was very unfortunate, and fpoiled all ; and he do's it, methinks, fo very awkwardly, that it is as difficult almoft to underftand his Language as his *Apes* ; if the Reader has a mind to attempt it, he will find it in the Margin (*p*).

Had *Albertus* only afferted, that the *Pygmies* were a fort of *Apes*, his Opinion poffibly might have obtained with lefs difficulty, unlefs he could have produced fome Body that had heard them talk. But *Ulyffes Aldrovandus* (*q*) is fo far from believing his *Ape Pygmies* ever fpoke, that he utterly denys, that there were ever any fuch Creatures in being, as the *Pygmies*, at all ; or that they ever fought the *Cranes*. *Cum itaque Pygmæos* (faith he) *dari negemus, Grues etiam cum iis Bellum gerere, ut fabulantur, negabimus, & tam pertinaciter id negabimus, ut ne juvantibus credemus.*

I find a great many very Learned Men are of this Opinion : And in the firft place, *Strabo* (*r*) is very pofitive; Ἐξαχῶς μὲν γ̀ ἐδεὶς ἐξηγεῖται τῶν πίστεως ἀξίων ἀνδεων i. e. *No Man worthy of belief did ever fee them.* And upon all occafions he declares the fame. So *Julius Cæfar Scaliger* (*s*) makes them to be only a Fiĉtion of the Ancients, *At hæc omnia* (faith he) *Antiquorum figmenta & meræ Nugæ, fi exftarent, reperirentur. At cum univerfus Orbis nunc nobis cognitus fit, nullibi hæc Naturæ Excrementa reperiri certiffimum eft.* And *Ifaac Cafaubon* (*t*) ridicules fuch as pretend to juftifie them : *Sic noftra ætate* (faith he) *non defunt, qui eandem de Pygmæis lepidam fabellam renovent ; ut qui etiam è Sacris Literis, fi Deo placet, fidem illis conentur aftruere. Legi etiam Bergei cujufdam Galli Scripta, qui fe vidiffe diceret. At non ego credulus illi, illi inquam Omnium Bipedum mendaciffimo.* I fhall add one Authority more, and that is of

---

(n) *Plinij Nat. Hift.* lib. 10. cap. 49.   (o) *Porphyrius de Abftinentia,* lib. 3.pag. m. 103.
(p) *Si qui Homines funt Silveftres, ficut Pygmeus, non fecundum unam rationem nobifcum diĉti funt Homines, fed aliquod habent Hominis in quadam deliberatione & Loquela,* &c.   A little after adds, *Voces quædam (fc.Animalia) formant ad diverfas conceptus quos habent, ficut Homo & Pygmæus ; & quedam non faciunt hoc, ficut multitudo fere tota aliorum Animalium. Adhuc autem eorum quæ ex ratione cogitativa formant voces, quædam funt fuccumbentia, quædam autem non fuccumbentia. Dico autem fuccumbentia, à conceptu Animæ cadentia & mota ad Naturæ Inftinĉtum, ficut Pygmeus, qui non, fequitur rationem Loquela fed Naturæ Inftinĉtum ; Homo autem non fuccumbit fed fequitur rationem.* Albert.Magn.de Animal.lib.t.cap.3. p.m.3.
(q) *Ulyf. Aldrovandi Ornitholog.* lib. 20. p. m. 344.   (r) *Strabo Geograph.* lib. 17. p. m. 565.
(s) *Jul. Cæf. Scaliger. Comment. in Arift. Hift. Animal.*lib.8. §. 126. p. m. 514.   (t) *Ifaac Caufabon Notæ & Cafligat. in* lib. 1. *Strabonis Geograph.* p. m. 38.

C 2   *Adrian*

Adrian Spigelius, who produces a Witness that had examined the very place, where the *Pygmies* were said to be ; yet upon a diligent enquiry, he could neither find them, nor hear any tidings of them. *Spigelius (u)* therefore tells us, *Hoc loco de Pygmæis dicendum erat , qui παρὰ πύγων᾽. dicti à statura, quæ ulnam non excedunt. Verùm ego Poetarum fabulas esse crediderim, pro quibus tamen* Aristoteles *minimè haberi vult, sed veram esse Historiam.* 8. Hist. Animal. 12. *asseverat. Ego quo minùs hoc statuam, tum Authoritate primùm Doctissimi* Strabonis 1. Geograph. *coactus sum, tum potissimùm nunc moveor, quod nostro tempore, quo nulla Mundi pars est, quam Nautarum Industria non perlustrarit , nihil tamen usquam simile aut visum est, aut auditum. Accedit quod* Franciscus Alvarez Lusitanus*, qui ea ipsa loca peragravit, circa quæ* Aristoteles *Pygmæos esse scribit, nullibi tamen tam parvam Gentem à se conspectam tradidit, sed Populum esse Mediocris staturæ, & Æthiopes tradit.*

I think my self therefore here obliged to make out, that there were such Creatures as *Pygmies*, before I determine what they were, since the very being of them is called in question, and utterly denied by so great Men, and by others too that might be here produced. Now in the doing this, *Aristotle*'s Assertion of them is so very positive, that I think there needs not a greater or better Proof; and it is so remarkable a one, that I find the very Enemies to this Opinion at a loss, how to shift it off. To lessen it's Authority they have interpolated the *Text*, by foisting into the *Translation* what is not in the Original ; or by not translating at all the most material passage, that makes against them ; or by miserably glossing it, to make him speak what he never intended : Such unfair dealings plainly argue, that at any rate they are willing to get rid of a Proof, that otherwise they can neither deny, or answer.

Aristotle's Text is this, which I shall give with *Theodorus Gaza*'s Translation : for discoursing of the Migration of Birds, according to the Season of the Year, from one Country to another, he saith (*w*) :

Μετὰ μὲν τὴν φθινοπωρινὴν Ἰσημερίαν,ἐκ τῶ Πόντε κὶ τῶν ψυχρῶν φεύγοντα τὸν ὅπείνετα χειμῶνα· μετὰ δὴ τὴν ἐαρινὴν, ἐκ τῶν θερινῶν, εἰς τὰς τόπες τὰς ψυχράς,φοβέμενα τὰ καύματα· τὰ μὲν, κὶ ἐκ τῶ ἐγγὺς τόπων ποιέμενα τὰς μεταβολὰς, τὰ ᾗ, κὶ ἐκ τῶν ἐχάταν,ὡς εἰπεῖν, οἷς αἱ γέρανοι ποιῦσι. Μεταβάλλυσι γὸ ἐκ τῶν Σκυθικῶν εἰς τὰ ἕλη τὰ ἄνω τῆς Αἰγύπτε,ὅθεν ὁ Νεῖλ. ῥεῖ. Ἔςι ᾗ ὁ τόπ. ἕτ. περὶ ὃν αἱ πυγμαῖ-	*Jam ab Antumnali Æquinoctio ex Ponto, Locisque frigidis fugiunt Hyemem futuram. A Verno autem ex tepida Regione ad frigidam sese conferunt, æstus metu futuri : & alia de locis vicinis discedunt , alia de ultimis, prope dixerim, ut Grues faciunt, quæ ex Scythicis Campis ad Paludes Ægypto superiores, unde Nilus profluit, veniunt, quo in loco pugnare cum Pygmæis dicuntur. Non enim id*

(u) *Adrian. Spigelij de Corporis Humani fabrica,* lib. 1. cap. 7. p. m. 15.   (w) *Aristotel. Hist.Animel.* lib. 8. cap. 12,

ei

οι κỳ]οικᾶσιν· ὃ γάρ ὅτι τᾶτο μᾶᾶ᾿Θ.,
ἀλλ᾿ ἔ῾ι κỳ]ὰ τῶ ἀλήθειαν. Γίν᾿Θ. μικρὸν μὲν, ὥσπερ λέγε]αι, ὴ αὐτοὶ ὴ οι
ἵππον· Τρωγλοδύτας δ᾿ εἰσὶ τὸν βίον.

fabula est, sed certè, genus tum ho-
minum, tum etiam Equorum pusillum
(ut dicitur) est, deguntque in Caver-
nis, unde Nomen Troglodytæ à sub-
eundis Cavernis accepere.

In Englifh 'tis thus: ' At the *Autumnal Æquinox* they go out of *Pontus*
' and the cold Countreys to avoid the Winter that is coming on. At the
' *Vernal Æquinox* they pafs from hot Countreys into cold ones,for fear of
' the enfuing Heat ; fome making their Migrations from nearer places ;
' others from the moft remote (as I'may fay) as the *Cranes* do : for they
' come out of *Scythia* to the Lakes above *Ægypt*,whence the *Nile* do's flow.
' This is the place, whereabout the *Pygmies* dwell: For this is no *Fable*,
' but a *Truth*. Both they and the Horfes, as 'tis faid, are a fmall kind.
' They are *Troglodytes*, or live in Caves.

We may here obferve how pofitive the *Philofopher* is, that there are
Pygmies ; he tells us where they dwell,and that 'tis no Fable,but a Truth.
But *Theodorus Gaza* has been unjuft in tranflating him, by foifting in,*Quo*
in loco pugnare cum Pygmæis dicuntur,whereas there is nothing in the Text
that warrants it : As likewife, where he expreffes the little Stature of the
Pygmies and the Horfes, there *Gaza* has rendered it, *Sed certè Genus tum*
Hominum, tum etiam Equorum pufillum. *Ariftotle* only faith, Γίν᾿Θ. μικρὸν
μὲν, ὥσπερ λέγε]αι, ὴ αὐτοὶ, ὴ οι ἵπποι. He neither makes his *Pygmies* Men,
nor faith any thing of their fighting the *Cranes* ; tho' here he had a fair
occafion, difcourfing of the Migration of the *Cranes* out of *Scythia* to the
Lakes above *Ægypt*,where he tells us the *Pygmies* are. *Cardan* (x) therefore muft certainly be out in his guefs, that *Ariftotle* only afferted the
Pygmies out of Complement to his Friend *Homer* ; for furely then he
would not have forgot their fight with the *Cranes* ; upon which occafion
only *Homer* mentions them (*). I fhould rather think that *Ariftotle*, being fenfible of the many Fables that had been raifed on this occafion,
ftudioufly avoided the mentioning this fight, that he might not give
countenance to the Extravagant Relations that had been made of it.

But I wonder that neither *Cafaubon* nor *Duvall* in their Editions of
Ariftotle's Works, fhould have taken notice of thefe Miftakes of *Gaza*,
and corrected them. And *Gefner*, and *Aldrovandus*, and feveral other
Learned Men, in quoting this place of *Ariftotle*, do make ufe of this
faulty Tranflation, which muft neceffarily lead them into Miftakes.
Sam. Bochartus (y) tho' he gives *Ariftotle*'s Text in Greek, and adds a new

(x) *Cardan de Rerum varietate*, lib.8. cap. 40. p. m. 153. (*) *Apparet ergo* (faith *Cardan*) *Pyg-*
mæorum Hiftoriam effe fabulofam, quod & Strabo fentit, & noftra ætas, cum omnia nunc fermè orbis mirabilia
innotuerint, declarat. Sed quod tantum Philofophum decepit, fuit Homeri Auctoritas non apud illem levis.
(y) *Bocharti Hierozoic. S. de Animalib. S. Script. part. Pofterior.* lib. 1. cap. 11. p. m. 76.

Tranflation

Translation of it, he leaves out indeed the *Cranes* fighting with the *Pyg-mies*,yet makes them *Men*,which *Aristotle* do's not ; and by anti-placing, *ut aiunt*, he renders *Aristotle's* Assertion more dubious ; *Neque enim* (saith he in the Translation) *id est fabula, sed reverâ, ut aiunt, Genus ibi par-vum est tam Hominum quàm Equorum.* *Julius Cæsar Scaliger* in transla-ting this Text of *Aristotle*, omits both these Interpretations of *Gaza* ; but on the other hand, is no less to be blamed in not translating at all the most remarkable passage, and where the *Philosopher* seems to be so much in earnest ; as, ὃ γὸ ἐςι τῶτο μῦθΘ., ἀλλ' ἐςι κατὰ τὼ ἀλήθειαν, this he leaves wholly out, without giving us his reason for it, if he had any : And *Scaliger's* (z) insinuation in his Comment , *viz.* *Negat esse fabulam de his (sc. Pygmeis)* Herodotus*, at Philosophus semper moderatus & pru-dens etiam addidit,* ὥσπερ λέγεται, is not to be allowed.   Nor can I as-sent to Sir *Thomas Brown's* (a) remark upon this place ; *Where indeed* (saith he) Aristotle *plays the* Aristotle ; *that is, the wary and evading As-sertor ; for tho' with* non est fabula *he seems at first to confirm it, yet at last he claps in,* sicut aiunt, *and shakes the belief he placed before upon it.   And therefore* Scaliger (saith he) *hath not translated the first, perhaps supposing it surreptitious, or unworthy so great an Assertor.* But had *Scaliger* known it to be surreptitious, no doubt but he would have remarked it ; and then there had been some Colour for the Gloss.   But 'tis unworthy to be believed of *Aristotle*, who was so wary and cautious, that he should in so short a passage, contradict himself ; and after he had so positively af-firmed the Truth of it, presently doubt it.   His ὥσπερ λέγεται therefore must have a Reference to what follows, *Pusillum genus, ut aiunt, ipsi atque etiam Equi,* as *Scaliger* himself translates it.

I do not here find *Aristotle* asserting or confirming any thing of the fa-bulous Narrations that had been made about the *Pygmies.*   He does not say that they were ἀνδρες, or ἀνθρωποι μικροὶ, or μέλανες ; he only calls them πυγμαῖοι.   And discoursing of the *Pygmies* in a place, where he is only treating about *Brutes*, 'tis reasonable to think, that he looked upon them only as such.   *This is the place where the* Pygmies *are ; this is no fable,* saith *Aristotle*, as 'tis that they are a Dwarfish Race of Men ; that they speak the *Indian* Language ; that they are excellent Archers ; that they are very Just ; and abundance of other Things that are fabulously reported of them ; and because he thought them *Fables*, he does not take the least notice of them , but only saith,*This is no Fable,but a Truth , that about the Lakes of* Nile *such Animals*, as are called *Pygmies*, do live. And, as if he had foreseen , that the abundance of Fables that *Ctesias* (whom he saith is not to be believed) and the *Indian Historians* had in-vented about them, would make the whole Story to appear as a Figment, and render it doubtful, whether there were ever such Creatures as *Pyg-*

(z) *Scalig... Comment. in Arist. Hist. Animal.* lib.8.p. m .914.   (b) Sir *Thomas Brown's Pseudodoxia, or, Enquiries into Vulgar Errors,* lib.4. cap. 11.

mics in Nature ; he more zealoufly afferts the *Being* of them, and affures us, That *this is no Fable, but a Truth.*

I fhall therefore now enquire what fort of Creatures thefe *Pygmies* were ; and hope, fo to manage the Matter, as in a great meafure, to a-bate the Paffion thefe Great Men have had againft them : for, no doubt, what has incenfed them the moft, was, the fabulous *Hiftorians* making them a part of *Mankind,* and then inventing a hundred ridiculous Sto-ries about them, which they would impofe upon the World as real Truths. If therefore they have Satisfaction given them in thefe two Points, I do not fee, but that the Bufinefs may be accommodated very fairly ; and that they may be allowed to be *Pygmies,* tho' we do not make them *Men.*

For I am not of *Gefner's* mind, *Sed veterum nullus* (faith he (b)) *ali-ter de Pygmæis fcripfit, quàm Homunciones effe.* Had they been a Race of *Men,* no doubt but *Ariftotle* would have informed himfelf farther about them. Such a Curiofity could not but have excited his Inquifitive *Ge-nius,* to a ftricter Enquiry and Examination ; and we might eafily have expected from him a larger Account of them. But finding them, it may be, a fort of *Apes,* he only tells us, that in fuch a place thefe *Pygmies* live.

Herodotus (c) plainly makes them *Brutes :* For reckoning up the *Ani-mals* of *Libya,* he tells us, Καὶ γὸ οἱ ὄφιες οἱ ὑπερμεγέθες, ἢ οἱ λέονlες καὶ ἃ τότες εἰσὶ, ἢ οἱ ἐλέφαντές τε ἢ ἄρκloι, ἢ ἀσπίδες τε ἢ ὄνοι οἱ τὰ κέρατα ἔχονlες ἢ οἱ κυνοκέφαλοι (in theMargin 'tis ἀκέφαλοι) οἱ ἐν τοῖσι στήθεσι τὰς ὀφθαλμὰς ἔχονlες (ὡς δὴ λέγεlαι γὸ ὑπὸ Λιϐύων) ἢ ἄγριοι ἄνδρες, ἢ γυναῖκες ἄγριαι ἢ ἄλλα πλήθεϊ πολλὰ θηρία ἀκαlάψδυσα i.e. *That there are here prodigious large Serpents, and Lions, and Elephants, and Bears, and Afps, and Affes that have horns, and Cyno-cephali,* (in the Margin 'tis *Acephali) that have Eyes in their Breaft , (as is reported by the Libyans) and wild Men, and wild Women, and a great ma-ny other wild Beafts that are not fabulous.* 'Tis evident therefore that *He-rodotus* his ἄγριοι ἄνδρες, ἢ γυναῖκες ἄγριαι are only θηρία or wild Beafts ; and tho' they are call'd ἄνδρες, they are no more *Men* than our *Orang-Outang,* or *Homo Sylveftris,* or *wild Man,* which has exactly the fame Name, and I muft confefs I can't but think is the fame Animal : and that the fame Name has been continued down to us, from his Time, and it may be from *Homer's.*

So *Philoftratus* fpeaking of *Æthiopia* and *Ægypt,* tells us (d), Βόσκεlαι δ ἢ θηρία οἷα ἐν ἑτέρωθι ἢ ἀνθρώπες μέλανας, ὁ μὴ ἄλλη ἤπειρος. Πυγμαί-ων τε ἐν αὐlαῖς ἔθνη, ἢ ὑλακlόνlων ἄλλο ἄλλη i.e. *Here are bred wild Beafts that are not in other places ; and black Men, which no other Country affords :*

(b) *Gefner. Hiftor. Quadruped.* p. m. 885.   (c) *Herodot. Melpomene feu* lib. 4. p. m. 285.   (d) *Phi-loftratus in vita Apollon. Tyanæi,* lib. 6. cap. 1. p. m. 258.

and

and amongst them is the Nation of the Pygmies, and the BARKERS, *that is,*
the *Cynocephali*.   For tho' *Philostratus* is pleafed here only to call them
Barkers, and to reckon them, as he does the *Black Men* and the *Pygmies*
amongft the *wild Beafts* of thofe Countreys ; yet *Ctefias*, from whom *Philoftratus* has borrowed a great deal of his *Natural Hiftory*, ftiles them
Men, and makes them fpeak, and to perform moft notable Feats in Mer-
chandifing.   But not being in a merry Humour it may be now, before
he was aware, he fpeaks Truth : For *Cælius Rhodiginus*'s (c) Character
of him is, *Philoftratus omnium qui unquam Hiftoriam confcripferunt, men-
daciffimus*.

Since the *Pygmies* therefore are fome of the *Brute Beafts* that naturally
breed in thefe Countries, and they are pleafed to let us know as much,
I can eafily excufe them a Name.   Ἄνδρες ἄγριοι, or *Orang-Outang*, is
alike to me ; and I am better pleafed with *Homer*'s ἄνδρες πυγμαῖοι, than
if he had called πίθηκοι.   Had this been the only Inftance where they had
mifapplied the Name of *Man*, methinks I could be fo good natur'd, as
in fome meafure to make an Apology for them.   But finding them fo
extravagantly loofe, fo wretchedly whimfical, in abufing the Dignity of
Mankind, by giving the Name of *Man* to fuch monftrous Productions
of their idle Imaginations, as the *Indian Hiftorians* have done, I do not
wonder that wife Men have fufpected all that comes out of their Mint, to
be falfe and counterfeit.

Such are their Ἀμύκτηρες or Ἄῤῥινες, that want Nofes, and have only
two holes above their Mouth ; they eat all things, but they muft be
raw ; they are fhort lived ; the upper part of their Mouths is very prominent.   The Ἐνωτοκεῖται, whofe Ears reach down to their Heels, on
which they lye and fleep.   The Ἄστομοι, that have no Mouths, a civil
fort of People, that dwell about the Head of the *Ganges* ; and live upon fmelling to boil'd Meats and the Odours of Fruits and Flowers ; they
can bear no ill fcent, and therefore can't live in a Camp.   The Μονόμ-
μαζοι or Μονόφθαλμοι, that have but one Eye, and that in the middle of
their Foreheads ; they have Dogs Ears ; their Hair ftands an end , but
fmooth on the Breafts.   The Στηνόφθαλμοι, that have Eyes in their
Breafts.   The Πάναι σφηνοκέφαλοι with Heads like Wedges.   The Μακρο-
κέφαλοι, with great Heads.   The Μαβρόρεοι, who live a Thoufand years.
The ἀκώποδες, fo fwift, that they will out-run a Horfe.   The ὀπισθοδάκ-
τυλοι, that go with their Heels forward, and their Toes backwards.   The
Μακρωσκελεῖς, The Στηανόποδες, The Μονοσκελεῖς, who have one Leg ,
but will jump a great way , and are call'd *Sciapodes*, becaufe when they
lye on their Backs , with this *Leg* they can keep off the Sun from their
Bodies.

(c) *Cælij Rhodigini Lection. Antiq.* lib. 17. cap. 13.

Now

Now *Strabo* (f), from whom I have collected the Description of these Monstrous sorts of *Men*, and they are mentioned too by *Pliny*, *Solinus*, *Mela*, *Philostratus*, and others; and *Munster* in his *Cosmography* (g) has given a *figure* of some of them; *Strabo*, I say, who was an Enemy to all such fabulous Relations, no doubt was prejudiced likewise against the *Pygmies*, because these *Historians* had made them a Puny Race of *Men*, and invented so many Romances about them. I can no ways therefore blame him for denying, that there were ever any such *Men Pygmies*; and do readily agree with him, that no *Man* ever saw them : and am so far from dissenting from those Great Men, who have denied them on this account, that I think they have all the reason in the World on their side. And to shew how ready I am to close with them in this Point, I will here examine the contrary Opinion, and what Reasons they give for the supporting it : For there have been some *Moderns*, as well as the *Ancients*, that have maintained that these *Pygmies* were real *Men*. And this they pretend to prove, both from *Humane Authority* and *Divine*.

Now by *Men Pygmies* we are by no means to understand *Dwarfs*. In all Countries, and in all Ages, there has been now and then observed such *Miniture* of Mankind, or under-sized Men. *Cardan* (h) tells us he saw one carried about in a Parrot's Cage, that was but a Cubit high. *Nicephorus* (i) tells us, that in *Theodosius* the Emperour's time, there was one in *Ægypt* that was no bigger than a Partridge; yet what was to be admired, he was very Prudent, had a sweet clear Voice, and a generous Mind; and lived Twenty Years. So likewise a King of *Portugal* sent to a Duke of *Savoy*, when he married his Daughter to him, an *Æthiopian Dwarf* but three Palms high (k). And *Thevenot* (l) tells us of the Present made by the King of the *Abyssins*, to the *Grand Seignior*, of several *little black Slaves* out of *Nubia*, and the Countries near *Æthiopia*, which being made *Eunuchs*, were to guard the Ladies of the *Seraglio*. And a great many such like Relations there are. But these being only *Dwarfs*, they must not be esteemed the *Pygmies* we are enquiring about, which are represented as a *Nation*, and the whole Race of them to be of the like stature. *Dari tamen integras Pumilionum Gentes, tam falsum est, quàm quod falsissimum*, saith *Harduin* (m).

Neither likewise must it be granted, that tho' in some *Climates* there might be *Men* generally of less stature, than what are to be met with in other Countries, that they are presently *Pygmies*. *Nature* has not fixed the same standard to the growth of *Mankind* in all Places alike, no more

---

(f) *Strabo Geograph.* lib.15.p.m.489. & lib.2.p.48. & alibi.　(g) *Munster Cosmograph.* lib.6.p.1151. (h) *Cardan de subtilitate*, lib. 11. p.458.　(i) *Nicephor. Hist. Ecclesiast.* lib.12. cap. 37.　(k) *Happelium in Relat. curiosis*, N°. 85. p. 677.　(l) *Thevenot. Voyage de Levant.* lib.2. c.68.　(m) *Jo. Harduini Nota in Plinij Nat. Hist.* lib. 6. cap. 22. p. 688.

D　　　　　　　　　　　　than

than to *Brutes* or *Plants*. The Dimensions of them all, according to the *Climate*, may differ. If we consult the Original, *viz. Homer* that first mentioned the *Pygmies*, there are only these two *Characteristics* he gives of them. That they are Πυγμαῖοι *seu Cubitales*; and that the *Cranes* did use to fight them. 'Tis true, as a *Poet*, he calls them ἄνδρες, which I have accounted for before. Now if there cannot be found such *Men* as are *Cubitales*, that the *Cranes* might probably fight with, notwithstanding all the *Romances* of the *Indian Historians*, I cannot think these *Pygmies* to be *Men*, but they must be some other *Animals*, or the whole must be a Fiction.

Having premised this, we will now enquire into their Assertion that maintain the *Pygmies* to be a Race of *Men*. Now because there have been *Giants* formerly, that have so much exceeded the usual Stature of *Man*, that there must be likewise *Pygmies* as defective in the other extream from this Standard, I think is no conclusive Argument, tho' made use of by some. Old *Caspar Bartholine* (*n*) tells us, that because *J. Cassanius* and others had wrote *de Gygantibus*, since no Body else had undertaken it, he would give us a Book *de Pygmæis*; and since he makes it his design to prove the Existence of *Pygmies*, and that the *Pygmies* were *Men*, I must confess I expected great Matters from him.

But I do not find he has informed us of any thing more of them, than what *Jo.Talentonius*, a Professor formerly at *Parma*, had told us before in his *Variarum & Reconditarum Rerum Thesaurus* (*o*), from whom he has borrowed most of this *Tract*. He has made it a little more formal indeed, by dividing it into *Chapters*; of which I will give you the *Titles*; and as I see occasion, some Remarks thereon : They will not be many, because I have prevented my self already. The *first Chapter* is, *De Homuncionibus & Pumilionibus seu Nanis à Pygmæis distinctis*. The *second Chapter*, *De Pygmæi nominibus & Etymologia*. The *third Chapter*, *Duplex esse Pygmæorum Genus; & primum Genus aliquando dari*. He means *Dwarfs*, that are no *Pygmies* at all. The *fourth Chapter* is, *Alterum Genus, nempe Gentem Pygmæorum esse, aut saltem aliquando fuisse Autoritatibus Humanis, fide tamen dignorum asseritur*. 'Tis as I find it printed; and no doubt an Error in the printing. The Authorities he gives, are, *Homer*, *Ctesias*, *Aristotle*, *Philostratus*, *Pliny*, *Juvenal*, *Oppian*, *Baptista Mantuan*, *St.Austin* and his *Scholiast. Ludovic. Vives*, *Jo. Laurentius Anania*, *Joh. Cassanius*, *Joh. Talentonius*, *Gellius*, *Pomp. Mela*, and *Olaus Magnus*. I have taken notice of most of them already, as I shall of St. *Austin* and *Ludovicus Vives* by and by. *Jo. Laurentius Anania* (*p*) *ex Mercatorum relatione tradit* (saith *Bartholine*) *eos* (*sc. Pygmæos*) *in Septentrionali Thraciæ Parte reperiri*, (*quæ Scythiæ est proxima*) *atque ibi cum Gruibus pugnare*. And *Joh. Cassanius* (*q*) (as he is here quoted) saith,

(n) *Caspar. Bartholin. Opusculum de Pygmæis*. (o) *Jo.Talentonij Variar.&c recondit. Rerum Thesaurus*, lib. 3. cap. 21. (p) *Joh. Laurent. Anania prope finem tractatus primi suæ Geograph.* (q) *Joh. Cassanius libello de Gygantibus*, p. 73.

De

De Pygmæis fabulosa quidem esse omnia, quæ de iis narrari solent, aliquando existimavi. Verùm cum videam non unum vel alterum, sed complures Classicos & probatos Autores de his Homunculis multa in eandem fere Sententiam tradidisse; eò adducor ut Pygmæos fuisse inficiari non ausim. He next brings in *Jo. Talentonius,* to whom he is so much beholden, and quotes his Opinion, which is full and home, *Constare arbitror* (saith *Talentonius*) (r) *debere concedi, Pygmæos non solùm olim fuisse, sed nunc etiam esse, & homines esse, nec parvitatem illis impedimento esse quo minùs sint & homines sint.* But were there such *Men Pygmies* now in being, no doubt but we must have heard of them; some or other of our Saylors, in their Voyages, would have lighted on them. Tho' *Aristotle* is here quoted, yet he does not make them *Men;* So neither does *Anania:* And I must own, tho' *Talentonius* be of this Opinion, yet he takes notice of the faulty Translation of this Text of *Aristotle* by *Gaza:* and tho' the parvity or lowness of Stature, be no Impediment, because we have frequently seen such *Dwarf-Men,* yet we did never see a *Nation* of them: For then there would be no need of that *Talmudical* Precept which *Job. Ludolphus* (s) mentions, *Nanus ne ducat Nanam, ne fortè oriatur ex iis Digitalis* (in *Bechor.* fol. 45.)

I had almost forgotten *Olaus Magnus,* whom *Bartholine* mentions in the close of this Chapter, but lays no great stress upon his Authority, because he tells us, he is fabulous in a great many other Relations, and he writes but by hear-say, that the *Greenlanders* fight the *Cranes; Tandem* (saith *Bartholine*) *neque ideo Pygmæi sunt, si fortè sagittis & hastis, sicut alij homines, Grues conficiunt & occidunt.* This I think is great Partiality: For *Ctesias,* an Author whom upon all turns *Bartholine* makes use of as an Evidence, is very positive, that the *Pygmies* were excellent *Archers:* so that he himself owns, that their being such, illustrates very much that *Text* in *Ezekiel,* on which he spends good part of the next Chapter, whose Title is, *Pygmæorum Gens ex Ezekiele, atque rationibus probabilibus adstruitur;* which we will consider by and by. And tho' *Olaus Magnus* may write some things by hear-say, yet he cannot be so fabulous as *Ctesias,* who (as *Lucian* tells us) writes what he neither saw himself, or heard from any Body else. Not that I think *Olaus Magnus* his *Greenlanders* were real *Pygmies,* no more than *Ctesias* his *Pygmies* were real *Men;* tho' he vouches very notably for them. And if all that have copied this Fable from *Ctesias,* must be look'd upon as the same Evidence with himself; the number of the *Testimonies* produced need not much concern us, since they must all stand or fall with him.

The *probable Reasons* that *Bartholine* gives in the *fifth Chapter,* are taken from other *Animals,* as Sheep, Oxen, Horses, Dogs, the *Indian Formica* and Plants: For observing in the same *Species* some excessive large,

---

(r) *Jo. Talentonius variar. & recondit. Rerum Thesaurus,* lib 3. cap. 21. p. m. 515.　(s) *Job. Ludolphi Comment. in Historiam Æthiopic.* p. m. 71.

and

and others extreamly little, he infers, *Quæ certè cum in Animalibus & Ve-*
getabilibus fiant ; cur in Humana specie non sit probabile, haud video : im-
primis cum detur magnitudinis excessus Gigantæus ; cur non etiam dabitur
Defectus ? Quia ergo dantur Gigantes, dabuntur & Pygmæi. Quam con-
sequentiam ut firmam, admittit Cardanus, (t) *licet de Pygmæis hoc tantùm*
concedat, qui pro miraculo, non pro Gente. Now *Cardan*, tho' he allows
this Confequence, yet in the fame place he gives feveral Reafons why
the *Pygmies* could not be *Men*, and looks upon the whole Story as fabulous. *Bartholine* concludes this *Chapter* thus : *Ulteriùs ut Probabilitatem*
fulciamus, addendum Sceleton Pygmæi, quod Drefdæ *vidimus inter alia plu-*
rima, servatum in Arce sereniss. Electoris Saxoniæ, *altitudine infra Cubitum,*
Ossium foliditate, proportioneque tum Capitis, tum aliorum ; ut Embrionem,
aut Artificiale quid Nemo rerum peritus suspicari possit. Addita insuper est
Inscriptio Veri Pygmæi. I hereupon looked into Dr. *Brown*'s Travels into thofe Parts, who has given us a large Catalogue of the Curiofities, the
Elector of *Saxony* had at *Drefden*, but did not find amongft them this
Sceleton ; which, by the largenefs of the Head, I fufpect to be the *Sce-*
leton of an *Orang-Outang*, or our *wild Man*. But had he given us either
a Figure of it, or a more particular Defcription, it had been a far greater
Satisfaction.

The Title of *Bartholine*'s *fixth Chapter* is, *Pygmæos esse aut fuisse ex va-*
riis eorum adjunctis, accidentibus , &c. *ab Authoribus descriptis ostenditur.*
As firft, their *Magnitude :* which he mentions from *Ctesias, Pliny, Gelli-*
as and *Juvenal ;* and tho' they do not all agree exactly, 'tis nothing.
Autorum hic dissensus nullus est (faith *Bartholine*) *etenim sicut in nostris ho-*
minibus, ita indubiè in Pygmæis non omnes ejusdem magnitudinis. 2.The
Place and *Country :* As *Ctesias* (he faith) places them in the middle of *In-*
dia ; Aristotle and *Pliny* at the Lakes above *Ægypt ; Homer*'s *Scholiast* in
the middle of *Ægypt ; Pliny* at another time faith they are at the Head
of the *Ganges*, and fometimes at *Gerania*, which is in *Thracia*, which being near *Scythia* , confirms (he faith) *Anania*'s *Relation. Mela* places
them at the *Arabian Gulf ;* and *Paulus Jovius docet Pygmæos ultra Japo-*
nem esse ; and adds, *has Autorum dissensiones facile fuerit conciliare ; nec*
mirum diversas relationes à Plinio *auditas.* For (faith he) as the *Tartars*
often change their Seats, fince they do not live in Houfes, but in Tents,
fo 'tis no wonder that the *Pygmies* often change theirs, fince inftead of
Houfes they live in Caves or Huts, built of Mud, Feathers , and Egg-
fhels. And this mutation of their Habitations he thinks is very plain
from *Pliny*, where fpeaking of *Gerania*, he faith, *Pygmæorum Gens* fuiffe
(*non jam esse*) *proditur, creduntque à Gruibus fugatos.* Which paffage
(faith *Bartholine*) had *Adrian Spigelius* confidered, he would not fo foon
have left *Aristotle*'s Opinion, becaufe *Franc. Alvares* the *Portuguefe* did
not find them in the place where *Aristotle* left them ; for the *Cranes*, it

may

may be, had driven them thence. His third Article is, their *Habitation*, which *Aristotle* saith is in *Caves*; hence they are *Troglodytes*. *Pliny* tells us they build Huts with Mud, Feathers, and Egg-shells. But what *Bartholine* adds, *Eò quod Terræ Cavernas inhabitent*, *non injurià dicti sunt olim Pygmæi, Terræ filii*, is wholly new to me, and I have not met with it in any Author before : tho' he gives us here several other significations of the word *Terræ filij* from a great many Authors, which I will not trouble you at present with. 4. The *Form*, being flat nosed and ugly, as *Ctesias*. 5. Their *Speech*, which was the same as the *Indians*, as *Ctesias*; and for this I find he has no other Author. 6. Their *Hair*; where he quotes *Ctesias* again, that they make use of it for *Clothes*. 7. Their *Vertues and Arts*; as that they use the same Laws as the *Indians*, are very just, excellent Archers, and that the King of *India* has Three thousand of them in his Guards. All from *Ctesias*. 8. Their *Animals*, as in *Ctesias*; and here are mentioned their Sheep, Oxen, Asses, Mules, and Horses. 9. Their various *Actions*; as what *Ctesias* relates of their killing Hares and Foxes with Crows, Eagles, &c. and fighting the *Cranes*, as *Homer*, *Pliny*, *Juvenal*.

The *seventh Chapter* in *Bartholine* has a promising Title, *An Pygmæi sint homines*, and I expected here something more to our purpose; but I find he rather endeavours to answer the Reasons of those that would make them *Apes*, than to lay down any of his own to prove them *Men*. And *Albertus Magnus*'s Opinion he thinks absurd, that makes them part Men part Beasts; they must be either one or the other, not a *Medium* between both; and to make out this, he gives us a large Quotation out of *Cardan*. But *Cardan* (*u*) in the same place argues that they are not Men. As to *Suessanus* (*w*) his Argument, that they want *Reason*, this he will not grant; but if they use it less, or more imperfectly than others (which yet, he saith, is not certain) by the same parity of Reason, *Children*, the *Bæotians*, *Cumani* and *Naturals* may not be reckoned *Men*; and he thinks, what he has mentioned in the preceding *Chapter* out of *Ctesias*, &c. shews that they have no small use of Reason. As to *Suessanus*'s next Argument, that they want Religion, Justice, &c. this, he saith, is not confirmed by any grave Writer; and if it was, yet it would not prove that they are not *Men*. For this defect (he saith) might hence happen, because they are forced to live in *Caves* for fear of the *Cranes*; and others besides them, are herein faulty. For this Opinion, that the *Pygmies* were *Apes* and not *Men*, he quotes likewise *Benedictus Varchius* (*x*), and *Joh. Tinnulus* (*y*), and *Paulus Jovius* (*z*), and several others of the Moderns, he tells us, are of the same mind. *Imprimis Geographici quos non puduit in Mappis Geographicis loco Pygmæorum simias cum Gruibus pugnantes ridiculè dipinxisse.*

(*u*) *Cardan. de Rerum varietate*, lib. 8. cap. 40.   (*w*) *Suessanus Comment. in Arist. de Histor. Animal.* lib. 8. cap. 12.   (*x*) *Benedict. Varchius de Monstris. lingua vernacula.*   (*y*) *Joh. Tinnulus in Glotto-Chrysio.*   (*z*) *Paulus Jovius lib. de Muscovit. Legatione.*

The

The Title of *Bartholine's eighth* and laſt Chapter is, *Argumenta eorum qui Pygmæorum Hiſtoriam fabuloſam cenſent, recitantur & refutantur.* Where he tells us, the only Perſon amongſt the Ancients that thought the Story of the *Pygmies* to be fabulous was *Strabo* ; but amongſt the Moderns there are ſeveral, as *Cardan, Budæus, Aldrovandus, Fullerus* and others. The firſt Objection (he ſaith) is that of *Spigelius* and others ; that ſince the whole World is now diſcovered, how happens it, that theſe *Pygmies* are not to be met with ? He has ſeven Anſwers to this Objection ; how ſatisfactory they are, the Reader may judge, if he pleaſes, by peruſing them amongſt the Quotations *(a)*. *Cardan's* ſecond Objection (he ſaith) is, that they live but eight years, whence ſeveral Inconveniences would happen, as *Cardan* ſhews ; he anſwers that no good Author aſſerts this ; and if there was, yet what *Cardan* urges would not follow ; and inſtances out of *Artemidorus* in *Pliny (b)*, as a *Parallel* in the *Calingæ* a Nation of *India*, where the *Women conceive when five years old, and do not live above eight.* *Geſner* ſpeaking of the *Pygmies*, ſaith, *Vitæ autem longitudo anni arciter octo ut* Albertus *refert.* *Cardan* perhaps had his Authority from *Albertus*, or it may be both took it from this paſſage in *Pliny*, which I think would better agree to *Apes* than *Men*. But *Artemidorus* being an *Indian Hiſtorian*, and in the ſame place telling other Romances, the leſs Credit is to be given to him. The third Objection, he ſaith, is of *Cornelius à Lapide*, who denies the *Pygmies*, becauſe *Homer* was the firſt Author of them. The fourth Objection he ſaith is, becauſe Authors differ about the Place where they ſhould be : This , he tells us, he has anſwered already in the fifth Chapter. The *fifth* and laſt Objection he mentions is, that but few have ſeen them. He anſwers, there are a great many Wonders in Sacred and Profane Hiſtory that we have not ſeen, yet muſt not deny. And he inſtances in three ; As the *Formicæ Indicæ*, which are as bigs as great Dogs: The *Cornu Plantabile* in the Iſland *Goa*, which when cut off from the Beaſt, and flung upon the Ground, will take root like a *Cabbage* : And the *Scotland Geeſe* that grow upon Trees, for which he quotes a great many Authors, and ſo concludes.

Now how far *Bartholine* in this Treatiſe has made out that the *Pygmies* of the Ancients were real *Men*, either from the Authorities he has quoted, or his Reaſonings upon them, I ſubmit to the Reader. I ſhall proceed now (as I promiſed) to conſider the Proof they pretend from *Ho-*

(a) *Reſpondeo* 1. *Contrarium teſtari Mercatorum Relationem apud Ananiam ſupra Cap.* 4. 2. *Et licet non inventi eſſent vivi à quolibet, pari jure Monocerota & alia negare liceret.* 3. *Qui maria pernavigant, vix oras paucas maritimas luſtrant, adeò non terras omnes à mari diſtitas.* 4. *Neque in Oris illos habitare maritimis ex Capite quinto manifeſtum eſt.* 5. *Qui teſtatum ſe omnem adhibuiſſe diligentiam in inquirendo eos ut invenitet.* 6 *Ita in terra habitant, ut in Antris vitam tolerare dicantur.* 7. *Si vel maximè omni ab omnibus diligentia quæſiti fuiſſent, nec inventi ; fieri poteſt, ut inſtar Gigantum jam deſierint nec ſint amplius.*
(b) *Plinij Hiſt. Nat.* lib. 7. cap. 2. p. m. 14.

ly Writ : For *Bartholine* and others infift upon that *Text* in *Ezekiel (Cap.* 27. *Verf.* 11.) where the *Vulgar* Tranflation has it thus ; *Filij Arvad cum Exercitu tuo fupra Muros tuos per circuitum ,* & *Pygmæi in Turribus tuis fuerunt* ; *Scuta fua fufpenderunt fupra Muros tuos per circuitum.* Now *Talentonius* and *Bartholine* think that what *Ctefias* relates of the *Pygmies,* as their being good *Archers,* very well illuftrates this Text of *Ezekiel :* I fhall here tranfcribe what Sir *Thomas Brown (c)* remarks upon it ; and if any one requires farther Satisfaction, they may confult *Job Ludolphus's Comment* on his *Æthiopic Hiftory (d).*

The fecond Teftimony (faith Sir *Thomas Brown) is deduced from Holy Scripture ; thus rendered in the Vulgar Tranflation,* Sed & Pygmæi qui e-rant in turribus tuis, pharetras fuas fufpenderunt in muris tuis per gyrum : *from whence notwithftanding we cannot infer this Affertion, for firft the Tranflators accord not, and the Hebrew word* Gammadim *is very varioufly rendered. Though* Aquila, Vatablus *and* Lyra *will have it* Pygmæi *, yet in the Septuagint, it is no more then Watchmen ; and fo in the* Arabick *and* High-Dutch. *In the* Chalde, Cappadocians, *in* Symmachus, Medes, *and in the* French, *thofe of* Gamed. Theodotian *of old, and* Tremellius *of late, have retained the Textuary word ; and fo have the* Italian, Low Dutch *and* Englifh *Tranflators, that is, the Men of* Arvad *were upon thy Walls round about, and the* Gammadims *were in thy Towers.*

Nor do Men only diffent in the Tranflation of the word , but in the Expofition of the Senfe and Meaning thereof ; for fome by Gammadims *underftand a People of* Syria, *fo called from the City of* Gamala ; *fome hereby underftand the* Cappadocians, *many the* Medes : *and hereof* Forerius *hath a fingular Expofition, conceiving the Watchmen of* Tyre, *might well be called* Pygmies, *the Towers of that City being fo high, that unto Men below, they appeared in a Cubital Stature. Others expound it quite contrary to common Acception, that is not Men of the leaft, but of the largeft fize ; fo doth* Cornelius *conftrue* Pygmæi, *or Viri Cubitales, that is, not Men of a Cubit high, but of the largeft Stature, whofe height like that of Giants, is rather to be taken by the Cubit than the Foot ; in which phrafe we read the meafure of* Goliah *, whofe height is faid to be fix Cubits and a fpan. Of affinity hereto is alfo the Expofition of* Jerom ; *not taking* Pygmies *for Dwarfs, but ftout and valiant Champions ; not taking the fenfe of* πυγμή, *which fignifies the Cubit meafure, but that which expreffeth* Pugils ; *that is, Men fit for Combat and the Exercife of the Fift. Thus can there be no fatisfying illation from this Text, the diverfity, or rather contrariety of Expofitions and Interpretations, diftracting more than confirming the Truth of the Story.*

But why *Aldrovandus* or *Cafpar Bartholine* fhould bring in St. *Auftin* as a Favourer of this Opinion of *Men Pygmies,* I fee no Reafon. To me

(c) Sir *Thomas Brown's Enquiries into Vulgar Errors,* lib. 4. cap. 11. p. 242.   (d) *Comment. in Hift. Æthiopic.* p. 73.

he

he feems to affert quite the contrary : For propofing this Queftion , *An ex propagine* Adam *vel filiorum* Noe, *quædam genera Hominum Monftrofa prodierunt ?* He mentions a great many monftrous Nations of *Men* , as they are defcribed by the *Indian Hiftorians*, and amongft the reft, the *Pygmies*, the *Sciopodes* , &c. And adds, *Quid dicam de* Cynocephalis, *quorum Canina Capita atque ipfe Latratus magis Beftias quàm Homines confitentur ? Sed omnia Genera Hominum, quæ dicuntur effe, effe credere, non eft neceffe.* And afterwards fo fully expreffes himfelf in favour of the *Hypothefis* I am here maintaining, that I think it a great Confirmation of it. *Nam & Simias* (faith he) *& Cercopithecos, & Sphingas, fi nefciremus non Homines effe, fed Beftias, poffent ifti Hiftorici de fua Curiofitate gloriantes velut Gentes aliquas Hominum nobis impunitâ vanitate mentiri.* At laft he concludes and determines the Queftion thus, *Aut illa , quæ talia de quibufdam Gentibus fcripta funt, omnino nulla funt, aut fi funt , Homines non funt, aut ex* Adam *funt fi homines funt.*

There is nothing therefore in St. *Auftin* that juftifies the being of *Men Pygmies*, or that the *Pygmies* were *Men* ; he rather makes them *Apes.* And there is nothing in his *Scholiaft Ludovicus Vives* that tends this way, he only quotes from other Authors, what might illuftrate the Text he is commenting upon, and no way afferts their being *Men.* I fhall therefore next enquire into *Bochartus*'s Opinion, who would have them to be the *Nubæ* or *Nobæ.* Hos Nubas Troglodyticos (faith (e) he) *ad A-valitem Sinum effe Pygmæos Veterum multa probant.* He gives us five Reafons to prove this. As , 1. The Authority of *Hefychius*, who faith Νᾶϐαι Πυγμαῖοι. 2. Becaufe *Homer* places the *Pygmies* near the Ocean, where the *Nubæ* were. 3. *Ariftotle* places them at the Lakes of the *Nile.* Now by the *Nile Bochartus* tells us, we muft underftand the *Aftaboras* , which the Ancients thought to be a Branch of the *Nile*, as he proves from *Pliny, Solinus* and *Æthicus.* And *Ptolomy* (he tells us) places the *Nubæ* hereabout 4. Becaufe *Ariftotle* makes the *Pygmies* to be *Troglodytes*, and fo were the *Nubæ.* 5. He urges that Story of *Nonnofus* which I have already mentioned, and thinks that thofe that *Nonnofus* met with, were a Colony of the *Nubæ* ; but afterwards adds , *Quos tamen abfit ut putemus Staturâ fuiffe Cubitali, prout Poetæ fingunt , qui omnia in majus augent.* But this methinks fpoils them from being *Pygmies* ; feveral other Nations at this rate may be *Pygmies* as well as thefe *Nubæ.* Befides, he does not inform us, that thefe *Nubæ* ufed to fight the *Cranes* ; and if they do not, and were not *Cubitales*, they can't be *Homer*'s *Pygmies*, which we are enquiring after. But the Notion of their being *Men*, had fo poffeffed him, that it put him upon fancying they muft be the *Nubæ* ; but 'tis plain that thofe in *Nonnofus* could not be a Colony of the *Nubæ* ; for then the *Nubæ* muft have underftood their Language, which the

(e) *Sam. Bochart. Geograph. Sacra*, Part. 1. lib. 2. cap. 23. p. m. 142.

Text

Text faith, none of the Neighbourhood did. And becaufe the *Nubæ* are *Troglodytes*, that therefore they muft be *Pygmies*, is no Argument at all. For *Troglodytes* here is ufed as an *Adjective* ; and there is a fort of *Sparrow* which is called *Paffer Troglodytes*. Not but that in *Africa* there was a Nation of *Men* called *Troglodytes*, but quite different from our *Pygmies*. How far *Bochartus* may be in the right, in guefling the Lakes of the *Nile* (whereabout *Ariftotle* places the *Pygmies*) to be the Fountains of the River *Aftaboras*, which in his defcription, and likewife the *Map*, he places in the Country of the *Avalie*, near the *Moffylon Emporium* ; I fhall not enquire. This I am certain of, he mifreprefents *Ariftotle* where he tells us *(f)*, *Quamvis in ea fabula hoc faltem verum effe afferat Philofophus, Pufillos Homines in iis locis degere :* for as I have already obferved ; *Ariftotle* in that *Text* faith nothing at all of their being *Men :* the contrary rather might be thence inferred, that they were *Brutes*. And *Bochart's* Tranflation, as well as *Gaza's* is faulty here, and by no means to be allowed, *viz. Ut aiunt, genus ibi parvum eft tam Hominum, quàm E-quorum;*which had *Bochartus* confidered, he would not have been fo fond it may be of his *Nubæ*. And if the Νοῦσοι Πυγμαῖοι in *Hefychius* are fuch *Pygmies* as *Bochartus* makes his *Nubæ; Quos tamen abfit ut putemus ftaturâ fuiffe Cubitali*, it will not do our bufinefs at all ; and neither *Homer's* Authority, nor *Ariftotle's* does him any Service.

But this Fable of *Men Pygmies* has not only obtained amongft the *Greeks* and *Indian Hiftorians :* the *Arabians* likewife tell much fuch Stories of them, as the fame learned *Bochartus* informs us. I will give his Latin Tranflation of one of them, which he has printed in *Arabick* alfo : *Arabes idem* (faith *(g) Bochartus) referunt ex cujufdam* Græculi *fide, qui* Jacobo Ifaaci *filio,* Sigarienfi *fertur ita narraffe. Navigabam aliquando in mari* Zingitano, *& impulit me ventus in quandam Infulam. In cujus Oppidum cum deveniffem, reperi Incolas Cubitalis effe ftaturæ, & plerofque Coclites. Quorum multitudo in me congregata me deduxit ad Regem fuum. Juffit is, ut Captivus detinerer ; & in quandam Caveæ fpeciem conjectus fum ; eos autem aliquando ad bellum inftrui cum viderem, dixerunt Hoftem imminere, & fore ut propediem ingrueret. Nec multò poft Gruum exercitus in eos infurrexit. Atque ideo erant Coclites, quod eorum oculos hæ confodiffent. Atque Ego, virgâ affumptâ, in eas impetum feci, & illæ avolârunt atque aufugerunt ; ob quod facinus in honore fui apud illos.* This Author, it feems, reprefents them under the fame Misfortune with the *Poet*, who firft mentioned them, as being blind, by having their Eyes peck'd out by their cruel Enemies. Such an Accident poffibly might happen now and then, in thefe bloody Engagements, tho' I wonder the *Indian Hiftorians* have not taken notice of it. However the *Pygmies* fhewed themfelves grateful to their Deliverer, in heaping *Honours* on him. One would guefs,

(f) *Bocharti Hierozoici pars Pofterior,* lib. 1. cap. 11. p. 76.   (g) *Bochartus ibid.* p. m. 77.

E   for

for their own fakes, they could not do lefs than make him their *Genera-liffimo* ; but our Author is modeft in not declaring what they were.

Isaac Voffius feems to unfettle all, and endeavours utterly to ruine the whole Story : for he tells us, If you travel all over *Africa*, you fhall not meet with either a *Crane* or *Pygmie* : *Se mirari* (faith *(h)* *Isaac Voffius*) Ariftotelem, *quod tam ferio affirmet non effe fabellam, quæ de Pygmæis & Bello, quod cum Gruibus gerant, narrantur.   Si quis totam pervadat* Afri-cam, *nullas vel Grues vel Pygmæos inveniet.*   Now one would wonder more at *Voffius*, that he fhould affert this of *Ariftotle*, which he never faid. And fince *Voffius* is fo miftaken in what he relates of *Ariftotle* ; where he might fo eafily have been in the right, 'tis not improbable, but he may be out in the reft too : For who has travelled all *Africa* over, that could inform him ? And why fhould he be fo peremptory in the Negative, when he had fo pofitive an Affirmation of *Ariftotle* to the contrary ? or if he would not believe *Ariftotle*'s Authority, methinks he fhould *Ariftophanes*'s, who tells us *(i)*, Σπείρειν ὅταν μὲν Γέρανθ. κρωίζων ἰς τὴν λιϐύηο μέταχωρῆ. *'Tis time to fow when the noify Cranes take their flight into* Libya.   Which Obfervation is likewife made by *Hefiod, Theognis, Aratus,* and others. And *Maximus Tyrius* (as I find him quoted in *Bochartus*) faith, Αἱ γέρα-νοι εἰς Αἰγύπῆε ὥρα Θέρυς ἀριςάμεναι, ἐκ ἀνεχόμεναι τὸ θάλπθ., τείνασαι πῆ-ρυχας ὥσπερ ἱςία, φέρονίαι διὰ τᾶ ἀερθ. εὐθὺ τῆ Σκυθῶν γῆς. i. e. *Grues per æ-ftatem ex* Ægypto *abfcedentes, quia Calorem pati non poffunt, alis velorum inftar expanfis, per aerem ad* Scythicam *plagam recta feruntur.* Which fully confirms that Migration of the *Cranes* that *Ariftotle* mentions.

But *Voffius* I find, tho' he will not allow the *Cranes*, yet upon fecond Thoughts did admit of *Pygmies* here : For this Story of the *Pygmies* and the *Cranes* having made fo much *noife*, he thinks there may be fome-thing of truth in it ; and then gives us his Conjecture, how that the *Pygmies* may be thofe *Dwarfs*, that are to be met with beyond the Foun-tains of the *Nile* ; but that they do not fight *Cranes* but *Elephants*, and kill a great many of them, and drive a confiderable Traffick for their Teeth with the *Jagi*, who fell them to thofe of *Congo* and the *Portuguefe*. I will give you *Voffius*'s own words ; *Attamen* (faith *(k)* he) *ut folent fabellæ non de nihilo fingi & aliquod plerunque continent veri, id ipfum quo-que hîc factum effe exiftimo.   Certum quippe eft ultra* Nili *fontes multos re-periri* Nanos, *qui tamen non cum Gruibus, fed cum Elephantis perpetuum ge-rant bellum.   Præcipuum quippe Eboris commercium in regno magni* Macoki *per iftos tranfigitur Homunciones ; habitant in Sylvis, & mira dexteritate Elephantos fagittis conficiunt.   Carnibus vefcuntur, Dentes verò* Jagis *diven-dunt, illi autem* Congentibus *&* Lufitanis.

(h) *Isaac Voffius de Nili aliorumque fluminum Origine*, Cap. 18.   (i) *Ariftophanes in Nubibus.*
(k) *Isaac Voffius ibid.*

Job Ludolphus (*l*) in his *Commentary* on his *Æthiopick History* remarks, That there was never known a Nation all of *Dwarfs*. *Nani quippe* (faith *Ludolphus*) *Naturæ quodam errore ex aliis justæ staturæ hominibus generantur. Qualis verò ea Gens sit, ex qua ista Naturæ Ludibria tantâ copiâ proveniant, Vossium docere oportebat, quia Pumiliònes Pumiles alios non gignunt, sed plerunque steriles sunt, experientia teste ; ut planè non opus habuerunt Doctores Talmudici Nanorum matrimonia prohibere, ne Digitales ex iis nascerentur.* *Ludolphus* it may be is a little too strict with *Vossius* for calling them *Nani* ; he may only mean a sort of Men in that Country of less Stature than ordinary. And *Dapper* in his History of *Africa*, from whom *Vossius* takes this Account, describes such in the Kingdom of *Mokoko*, he calls *Mimos*, and tells us that they kill *Elephants*. But I see no reason why *Vossius* should take these Men for the *Pygmies* of the Ancients, or think that they gave any occasion or ground for the inventing this Fable, if there was no other reason, this was sufficient, because they were able to kill the *Elephants*. The *Pygmies* were scarce a Match for the *Cranes* ; and for them to have encountred an *Elephant*, were as vain an Attempt, as the *Pygmies* were guilty of in *Philostratus* (*m*), 'who to re-' venge the Death of *Antæus*, having found *Hercules* napping in *Libya*, ' mustered up all their Forces against him. One *Phalanx* (he tells us) af-' saulted his left hand ; but against his right hand, that being the stron-' ger, two *Phalanges* were appointed. The Archers and Slingers besieg-' ed his feet admiring the hugeness of his Thighs : But against his Head, ' as the Arsenal, they raised Batteries, the King himself taking his Post ' there. They set fire to his Hair, put Reaping-hooks in his Eyes ; and ' that he might not breath , clapp'd Doors to his Mouth and Nostrils ; ' but all the Execution that they could do, was only to awake him, ' which when done, deriding their folly, he gather'd them all up into his ' Lion's Skin, and carried them (*Philostratus* thinks) to *Euristhenes*. This *Antæus* was as remarkable for his height, as the *Pygmies* were for their lowness of Stature : For *Plutarch* (*n*) tells us, that *Q. Sertorius* not being willing to trust Common Fame, when he came to *Tingis* (now *Tangier*) he caused *Antæus*'s Sepulchre to be opened , and found his Corps full threescore Cubits long. But *Sertorius* knew well enough how to impose upon the Credulity of the People, as is evident from the Story of his *white Hind*, which *Plutarch* likewise relates.

But to return to our *Pygmies* ; tho' most of the great and learned Men would seem to decry this Story as a Fiction and meer Fable, yet there is something of Truth, they think, must have given the first rise to it, and that it was not wholly the product of Phancy, but had some real foundation, tho' disguised , according to the different Imagination and *Genius*

---

(1) *Job Ludolphus in Comment. in Historiam Æthiopicam*, p. m. 71.   (m) *Philostratus. Icon.* lib. 2. p. m. 817.   (n) *Plutarch. in vita Q. Sertorij.*

E 2   of

of the *Relator :* 'Tis this that has incited them to give their several Con-
jectures about it.   *Job Ludolphus* finding what has been offered at in Relation to the *Pygmies,* not to satisfie, he thinks he can better account for
this Story, by leaving out the *Cranes,* and placing in their stead, another
sort of Bird he calls the *Condor.*   I will give you his own words: *Sed
ad Pygmæos* (faith (*o*) *Ludolphus*) *revertamur ; fabula de Geranomachia
Pygmæorum seu pugna cum Gruibus etiam aliquid de vero trahere videtur, ſi
pro Gruibus* Condoras *intelligas, Aves In interiore* Africa *maximas, ut ſi-
dem penè excedat ; aiunt enim quod Ales iſta vitulum Elephanti in Aerem
extollere poſſit ; ut infra docebimus.   Cum his Pygmæos pugnare , ne pecora
ſua rapiant , incredibile non eſt.   Error ex eo natus videtur, quod primus
Relator, alio vocabulo deſtitutus,* Grues *pro* Condoris *nominârit, ſicuti* Plau-
tus *Picos pro Gryphibus, &* Romani Boves lucas *pro Elephantis dixere.*

   'Tis true, if what *Juvenal* only in ridicule mentions, was to be admitted as a thing really done, that the *Cranes* could fly away with a *Pygmie,* as our *Kites* can with a Chicken, there might be some pretence for
Ludovicus's *Condor* or *Cunctor :* For he mentions afterwards (*p*) out of
P. *Joh. dos Santos* the *Portugneſe,* that 'twas observed that one of these
Condors once flew away with an Ape, Chain , Clog and all, about
ten or twelve pounds weight, which he carried to a neighbouring Wood,
and there devoured him.   And *Garcilaſſo de la Vega* (*q*) relates that
they will seize and fly away with a Child ten or twelve years old.
But *Juvenal* (*r*) only mentions this in ridicule and merriment , where
he saith,

> *Ad ſubitas Thracum volucres, nubemque ſonoram
> Pygmæos parvis currit Bellator in armis :
> Mox impar hoſti, raptuſque per acra curvis
> Unguibus à ſævâ fertur Grue.*

   Besides, were the *Condors* to be taken for the *Cranes,* it would utterly
spoil the *Pygmæomachia ;* for where the Match is so very unequal , 'tis
impoſſible for the *Pygmies* to make the least ſhew of a fight.   *Ludolphus*
puts as great hardſhips on them, to fight these *Condors ,* as *Voſſius* did, in
making them fight *Elephants,*but not with equal Succeſs; for *Voſſius*'s *Pygmies* made great Slaughters of the *Elephants ;* but *Ludolphus* his *Cranes*
sweep away the *Pygmies,*as easily as an *Owl* would a *Mouſe,* and eat them
up into the bargain ;   now I never heard the *Cranes* were so cruel and
barbarous to their Enemies, tho' there are some Nations in the World
that are reported to do so.

   Moreover, these *Condor*'s I find are very rare to be met with ; and

   (o) *Job Ludolphus Comment. in Hiſtoriam ſuam Æthiopic.* p. 73.   (p) *Job Ludolphus* ibid. pag. 264.
   (q) *Garcilaſſo de la Vega Royal Comment. of* Peru.   (r) *Juvenal Satyr.* 13. verſ. 167.

when

when they are, they often appear fingle, or but a few. Now *Homer*'s, and the *Cranes* of the Ancients, are always reprefented in Flocks. Thus *Oppian (s)* as I find him tranflated into Latin Verfe :

Et velut Æthiopum veniunt, Nilique fluenta
Turmatim Palamedis Aves, celfæque per altum
Aera labentes fugiunt Athlanta nivofum,
Pygmæos imbelle Genus, parvumque fatigant,
Non perturbato procedunt ordine denfæ
Inftructis volucres obfcurant aëra Turmis.

To imagine thefe *Grues* a fingle Gigantick Bird, would much leffen the Beauty of *Homer*'s *Simile*, and would not have ferved his turn ; and there are none who have borrowed *Homer*'s fancy, but have thought fo. I will only farther inftance in *Baptifta Mantuan :*

Pygmæi breve vulgus, iners Plebecula, quando
Convenere Grues longis in prælia roftris,
Sublato clamore fremunt, dumque agmine magno
Hoftibus occurrit, tellus tremit Indica, clamant
Littora, arenarum nimbis abfconditur aër ;
Omnis & involuit Pulvis folemque, Polumque,
Et Genus hoc Hominum naturâ imbelle, quietum,
Mite, facit Mavors pugnax, immane Cruentum.

Having now confidered and examined the various Opinions of thefe learned Men concerning this *Pygmæomachia* ; and reprefented the Reafons they give for maintaining their Conjectures ; I fhall beg leave to fubjoyn my own : and if what at prefent I offer, may feem more probable, or account for this Story with more likelyhood, than what hath hitherto been advanced, I fhall not think my time altogether mifpent : But if this will not do, I fhall never trouble my head more about them , nor think my felf any ways concerned to write on this Argument again. And I had not done it now, but upon the occafion of Diffecting this *Orang-Outang*, or *wild Man*, which being a Native of *Africa*, and brought from *Angola*, tho' firft taken higher up in the Country , as I was informed by the Relation given me ; and obferving fo great a Refemblance, both in the outward fhape, and, what furprized me more, in the Structure likewife of the inward Parts, to a *Man* ; this Thought was eafily fuggefted to me, That very probably this *Animal*, or fome other fuch of the fame *Species*, might give the firft rife and occafion to the Stories of the *Pygmies*. What has been the πρῶτον ψευδὸς, and rendered this Story fo difficult to be believed, I find hath been the Opinion that has generally

(s) Oppian. lib. 1. de Pifcibus.

obtained,

obtained, that these *Pygmies* were really a Race of *little Men*. And tho' they are only *Brutes*, yet being at first call'd *wild Men*, no doubt from the Resemblance they bear to *Men* ; there have not been wanting those especially amongst the Ancients, who have invented a hundred ridiculous Stories concerning them ; and have attributed those things to them, were they to be believed in what they say, that necessarily conclude them real *Men*.

To sum up therefore what I have already discoursed, I think I have proved, that the *Pygmies* were not an *Humane Species* or *Men*. And tho' *Homer*, who first mentioned them, calls them ἄνδρες πυγμαῖοι, yet we need not understand by this Expression any thing more than *Apes* : And tho' his *Geranomachia* hath been look'd upon by most only as a Poetical Fiction ; yet by assigning what might be the true Cause of this Quarrel between the *Cranes* and *Pygmies*, and by divesting it of the many fabulous Relations that the *Indian Historians*, and others, have loaded it with, I have endeavoured to render it a true, at least a probable Story. I have instanced in *Ctesias* and the *Indian Historians*, as the Authors and Inventors of the many Fables we have had concerning them : Particularly, I have Examined those Relations, where Speech or Language ·is attributed to them ; and shewn, that there is no reason to believe, that they ever spake any Language at all. But these *Indian Historians* having related so many extravagant Romances of the *Pygmies*, as to render their whole History suspected, nay to be utterly denied, that there were ever any such Creatures as *Pygmies* in *Nature*, both by *Strabo* of old, and most of our Learned Men of late, I have endeavoured to assert the Truth of their *being*, from a *Text* in *Aristotle* ; which being so positive in affirming their Existence, creates a difficulty, that can no ways be got over by such as are of the contrary Opinion. This *Text* I have vindicated from the false Interpretations and Glosses of several Great Men, who had their Minds so prepossessed and prejudiced with the Notion of *Men Pygmies*, that they often would quote it, and misapply it, tho' it contain'd nothing that any ways favoured their Opinion ; but the contrary rather, that they were *Brutes*, and not *Men*.

And that the *Pygmies* were really *Brutes*, I think I have plainly proved out of *Herodotus* and *Philostratus*, who reckon them amongst the *wild Beasts* that breed in those Countries : For tho' by *Herodotus* they are call'd ἄνδρες ἄγριοι, and *Philostratus* calls them ἄνθρωπες μέλανας, yet both make them θηρία or *wild Beasts*. And I might here add what *Pausanias* (t) relates from *Euphemus Car*, who by contrary Winds was driven upon some Islands, where he tells us, ἐν ᾗ ταύταις οἰκεῖν ἄνδρας ἀγρίυς, but when he comes to describe them, tells us that they had no Speech ;

(t) *Pausanias in Atticis*, p. m. 21.

that they had Tails on their Rumps; and were very lafcivious toward the Women in the Ship. But of thefe more, when we come to difcourfe of *Satyrs*.

And we may the lefs wonder to find that they call *Brutes Men*, fince 'twas common for thefe *Hiftorians* to give the Title of *Men*, not only to *Brutes*, but they were grown fo wanton in their Inventions, as to defcribe feveral Nations of *Monftrous Men*, that had never any Being, but in their own Imagination, as I have inftanced in feveral. I therefore excufe *Strabo* for denying the *Pygmies*, fince he could not but be convinced, they could not be fuch *Men*, as thefe *Hiftorians* have defcribed them. And the better to judge of the Reafons that fome of the Moderns have given to prove the Being of *Men Pygmies*, I have laid down as *Poftulata's*, that hereby we muft not underftand *Dwarfs*, nor yet a Nation of *Men*, tho' fomewhat of a leffer fize and ftature than ordinary; but we muft obferve thofe two Charaƈterifticks that *Homer* gives of them, that they are *Cubitales*, and fight *Cranes*.

Having premifed this, I have taken into confideration *Cafpar Bartholine Senior* his *Opufculum de Pygmæis*, and *Jo. Talentonius's* Differtation about them; and upon examination do find, that neither the Humane Authorities, nor Divine that they alledge, do any ways prove, as they pretend, the Being of *Men Pygmies*. St. *Auftin*, who is likewife quoted on their fide, is fo far from favouring this Opinion, that he doubts whether any fuch Creatures exift, and if they do, concludes them to be *Apes* or *Monkeys*; and cenfures thofe *Indian Hiftorians* for impofing fuch Beafts upon us, as diftinƈt Races of *Men*. *Julius Cæfar Scaliger*, and *Ifaac Cafaubon*, and *Adrian Spigelius* utterly deny the Being of *Pygmies*, and look upon them as a Figment only of the Ancients, becaufe fuch little Men as they defcribe them to be, are no where to be met with in all the World. The Learned *Bochartus*, tho' he efteems the *Geranomachia* to be a Fable, and flights it, yet thinks that what might give the occafion to the Story of the *Pygmies*, might be the *Nubæ* or *Nobæ*; as *Ifaac Voffius* conjeƈtures that it was thofe *Dwarfs* beyond the Fountains of the *Nile*, that *Dapper* calls the *Mimos*, and tells us, they kill *Elephants* for to make a Traffick with their Teeth. But *Job Ludolphus* alters the Scene, and inftead of *Cranes*, fubftitutes his *Condors*, who do not fight the *Pygmies*, but fly away with them, and then devour them.

Now all thefe Conjeƈtures do no ways account for *Homer's Pygmies* and *Cranes*, they are too much forced and ftrain'd. Truth is always eafie and plain. In our prefent Cafe therefore I think the *Orang-Outang*, or *wild Man*, may exaƈtly fupply the place of the *Pygmies*, and without any violence or injury to the Story, fufficiently account for the whole Hiftory of the *Pygmies*, but what is moft apparently fabulous; for what has been the greateft difficulty to be folved or fatisfied, was their being
Men,

Men ; for as *Gesner* remarks (as I have already quoted him) *Sed vete-rum nullus aliter de Pygmæis scripsit,quàm Homunciones esse.* And the Mo-derns too, being byassed and misguided by this Notion, have either wholly denied them, or contented themselves in offering their Conje-ctures what might give the first rise to the inventing this Fable. And tho' *Albertus*, as I find him frequently quoted, thought that the *Pygmies* might be only a sort of *Apes*, and he is placed in the Head of those that espoused this Opinion, yet he spoils all, by his way of reasoning, and by making them speak; which was more than he needed to do.

I cannot see therefore any thing that will so fairly solve this doubt, that will reconcile all, that will so easily and plainly make out this Sto-ry, as by making the *Orang-Outang* to be the *Pygmie* of the Ancients; for 'tis the same Name that Antiquity gave them. For *Herodotus's* ἄνδρες ἄγριοι, what can they be else, than *Homines Sylvestres*, or *wild Men ?* as they are now called. And *Homer's* ἄνδρες πυγμαῖοι , are no more an Humane Kind, or Men, than *Herodotus's* ἄνδρες ἄγριοι, which he makes to be Θηρία, or *wild Beasts :* And the ἄνδρες μικροὶ or μέλανες (as they are often called) were just the same. Because this sort of *Apes* had so great a resemblance to Men,more than other *Apes* or *Monkeys*; and they going naturally erect,and being designed by Nature to go so, (as I have shewn in the *Anatomy*) the Ancients had a very plausible ground for giving them this denomination of ἄνδρες or ἄνθρωποι , but commonly they added an Epithet ; as ἄγριοι , μικροὶ , πυγμαῖοι , μέλανες , or some such like. Now the Ancient *Greek* and *Indian Historians* , tho' they might know these *Pygmies* to be only *Apes* like Men, and not to be real *Men*, yet being so extreamly addicted to *Mythology*, or making Fables, and finding this so fit a Subject to engraft upon, and invent Sto-ries about, they have not been wanting in furnishing us with a great many very Romantick ones on this occasion. And the Moderns being imposed upon by them, and misguided by the Name of ἄνδρες or ἄν-θρωποι, as if thereby must be always understood an *Humane Kind*, or *re-al Men*, they have altogether mistaken the Truth of the Story, and have either wholly denied it, or rendered it as improbable by their own Con-jectures.

This difficulty therefore of their being called *Men*, I think, may fair-ly enough be accounted for by what I have said. But it may be object-ed that the *Orang-Outang*, or these *wild* or *savage Men* are not πυγμαῖοι, or *Trispithami*, that is, but two Foot and a quarter high,because by some Relations that have been given, it appears they have been observed to be of a higher stature, and as tall as ordinary Men. Now tho' this may be allowed as to these *wild Men* that are bred in other places; and pro-bably enough likewise, there are such in some Parts of the Continent of *Africa* ; yet 'tis sufficient to our business if there are any there, that will come within our Dimensions; for our Scene lies in *Africa* ; where *Strabo* observes, that generally the Beasts are of a less size than ordinary ; and

this

this he thinks might give the rife to the Story of the *Pygmies*. For, faith he (*u*), Τὰ ἢ βοσκήματα αὐτοῖς ἐςὶ μικρὰ, πρόβατα ἢ αἶγες, ἢ κύνες μικροὶ, τραχεῖς ἢ ἢ μάχιμοι (οἰκοῦντες μικροὶ ὄντες) τάχα ἢ ἢ τὰς πυγμαίας ἀπὸ ℸ τούτων μικρότητας ἐπενόησαν, ἢ ἀνέπλασαν. i.e. *That their Beafts are fmall, as their Sheep, Goats and Oxen, and their Dogs are fmall, but hairy and fierce: and it may be* (faith he) *from the* μικρότητα *or littlenefs of the ftature of thefe Animals, they have invented and impofed on us the* Pygmies. And then adds, *That no body fit to be believed ever faw them*; becaufe he fancied, as a great many others have done, that thefe *Pygmies* muft be *real Men*, and not a fort of *Brutes*. Now fince the other *Brutes* in this Country are generally of a lefs fize than in other Parts, why may not this fort of *Ape*, the *Orang-Outang*, or *wild Man*, be fo likewife. *Ariftotle* fpeaking of the *Pygmies*, faith, γένϕ μικρὸν μὲν ἢ αὐτοὶ, ἢ οἱ ἵπποι *That both they and the Horfes there are but fmall.* He does not fay *their Horfes*, for they were never mounted upon *Horfes*, but only upon *Partridges, Goats* and *Rams*. And as the *Horfes*, and other *Beafts* are naturally lefs in *Africa* than in other Parts, fo likewife may the *Orang-Outang* be. This that I diffected, which was brought from *Angola* (as I have often mentioned) wanted fomething of the juft ftature of the *Pygmies*; but it was young, and I am therefore uncertain to what tallnefs it might grow, when at full Age: And neither *Tulpius*, nor *Gaffendus*, nor any that I have hitherto met with, have adjufted the full ftature of this *Animal* that is found in thofe Parts from whence ours was brought: But 'tis moft certain, that there are forts of *Apes* that are much lefs than the *Pygmies* are defcribed to be. And, as other *Brutes*, fo the *Ape-kind*, in different Climates, may be of different Dimenfions; and becaufe the other *Brutes* here are generally fmall, why may not *they* be fo likewife. Or if the difference fhould be but little, I fee no great reafon in this cafe, why we fhould be over-nice, or fcrupulous.

As to our *Ape Pygmies* or *Orang-Outang* fighting the *Cranes*, this, I think, may be eafily enough made out, by what I have already obferved; for this *wild Man* I diffected was *Carnivorous*, and it may be *Omnivorous*, at leaft as much as *Man* is; for it would eat any thing that was brought to the Table. And if it was not their Hunger that drove them to it, their Wantonnefs, it may be, would make them apt enough to rob the *Cranes* Nefts; and if they did fo, no doubt but the *Cranes* would make noife enough about it, and endeavour what they could to beat them off, which a Poet might eafily make a Fight: Tho' *Homer* only makes ufe of it, as a *Simile*, in comparing the great Shouts of the *Trojans* to the Noife of the *Cranes*, and the Silence of the *Greeks* to that of the *Pygmies* when they are going to Engage, which is natural enough, and very juft, and contains nothing, but what may eafily be believed; tho' upon this account he is commonly exposed, and derided, as the Inventor of this Fable; and that there was nothing of Truth in it, but that 'twas wholly a Fiction of his own.

Those *Pygmies* that *Paulus Jovius (w)* describes, tho' they dwell at a great distance from *Africa*, and he calls them *Men*, yet are so like *Apes*, that I cannot think them any thing else. I will give you his own words : *Ultra* Lapones (saith he) *in Regione inter* Corum *&* Aquilonem *perpetuâ oppreſſa Caligine* Pygmæos *reperiri, aliqui eximiæ fidei teſtes retulerunt ; qui poſtquàm ad ſummum adoleverint, noſtratis Pueri denum annorum Menſuram vix excedunt. Meticuloſum genus hominum, & garritu Sermonem exprimens, adeo ut tam Simiæ propinqui , quàm ſtaturâ ac ſenſibus ab juſtæ Proceritatis homine remoti videantur.* Now there is this Advantage in our *Hypotheſis*, it will take in all the *Pygmies*, in any part of the World, or wherever they are to be met with, without ſuppoſing, as ſome have done, that 'twas the *Cranes* that forced them to quit their Quarters ; and upon this account ſeveral Authors have deſcribed them in different places : For unleſs we ſuppoſe the *Cranes* ſo kind to them, as to waft them over, how came we to find them often in Iſlands ? But this is more than can be reaſonably expected from ſo great Enemies.

I ſhall conclude by obſerving to you, that this having been the Common Error of the Age, in believing the *Pygmies* to be a ſort of *little Men* , and it having been handed down from ſo great Antiquity, what might contribute farther to the confirming this Miſtake , might be, the Impoſture of the Navigators, who ſailing to theſe Parts where theſe *Apes* are, they have embalmed their Bodies, and brought them home, and then made the People believe that they were the *Men* of thoſe Countries from whence they came. This *M. P. Venetus* aſſures us to have been done ; and 'tis not unlikely : For, ſaith he (†), *Abundat quoque Regio ipſa (ſc.* Baſman in Java majori*) diverſis Simiis magnis & parvis, hominibus ſimillimis, hos capiunt Venatores & totos depilant, niſi quòd in barba & in loco ſecreto Pilos relinquunt , & occiſos ſpeciebus Aromaticis condiunt , & poſtea deſiccant, venduntque Negociatoribus , qui per diverſas Orbis Partes Corpora illa deferentes, homines perſuadent Tales Homunciones in Maris Inſulis reperiri.* *Joh. Jonſton (x)* relates the ſame thing , but without quoting the Author; and as he is very apt to do, commits a great miſtake, in telling us, *pro Homunculis marinis venditant.*

I ſhall only add, That the Servile Offices that theſe Creatures are obſerved to perform, might formerly, as it does to this very day, impoſe upon Mankind to believe, that they were of the ſame *Species* with them-

(w) *Paul. Jovij de Legatione Muſchovitar.* lib.p. m.489.   (†) *M.Pauli Veneti de Regionibus Oriental.* Lib. 3. cap. 15. p. m. 390.   (x) *Jo. Jonſton. Hiſt. Nat. de Quadruped.* p. m. 139.

felves ; but that only out of fullennefs or cunning, they think they will
not *fpeak*, for fear of being made Slaves. *Philoftratus* (*y*) tells us, That
the *Indians* make ufe of the *Apes* in gathering the Pepper ; and for this
Reafon they do defend and preferve them from the *Lions* , who are very greedy of preying upon them : And altho' he calls them *Apes*, yet
he fpeaks of them as *Men*, and as if they were the Husbandmen of the
Pepper Trees, ὡ τὰ δὶνδερ αἱ πιπερίδες, ὧν γεωργοὶ πίθηκοι. And he calls
them the People of *Apes* ; ἃ λέγεται πιθήκων οἰκεῖν δῆμος ἐν μυχοῖς τοῦ ὄρεος.
Dapper (*z*) tells us, That the Indians *take the* Baris *when young, and
make them fo tame, that they will do almoft the work of a Slave* ; *for they
commonly go erect as Men do*. *They will beat Rice in a Mortar* , *carry
Water in a Pitcher*, &c. And *Gaffendus* (*a*) in the Life of *Pieresky*,
tells us, *That they will play upon a Pipe or Cittern*, *or the like Mufick* , *they
will fweep the Houfe, turn the Spit, beat in a Mortar* , *and do other Offices
in a Family*. And *Acofta*, as I find him quoted by *Garcilaffo de la Vega*
(*b*) tells us of a *Monkey* he faw at the Governour's Houfe at *Cartagena*,
' whom they fent often to the Tavern for Wine, with Money in one
' hand, and a Bottle in the other ; and that when he came to the Ta-
' vern, he would not deliver his Money , until he had received his
' Wine. If the Boys met with him by the way, or made a houting
' or noife after him, he would fet down his Bottle, and throw Stones
' at them ; and having cleared the way , he would take up his Bottle,
' and haften home. And tho' he loved Wine exceffively, yet he would
' not dare to touch it, unlefs his Mafter gave him Licenfe. A great many Inftances of this Nature might be given that are very furprifing. And
in another place tells us, That the Natives think that they can fpeak,
but will not, for fear of being made to work. And *Bontius* (*c*) mentions that the *Javans* had the fame Opinion concerning the *Orang-
Outang*, *Loqui verò eos, eafque Javani aiunt, fed non velle, ne ad labores
cogerentur*.

(y) *Philoftratus in vita Apollonij Tyanæi*, lib. 3. cap. 1. p. m. 110, & 111.   (z) *Dapper Defcription
de l'Afrique*, p. m. 249.   (a) *Gaffendus in vita Pierskij*, lib. 5. p. m. 169.   (b) *Garcilaffo de la Vega Royal Commentaries* of *Peru*, lib. 8. cap. 18. p. 1333.   (c) *Jac. Bontij Hift. Nat. & Med.* lib. 5.
cap. 32. p. m. 85.

A
Philological Essay
Concerning the
CYNOCEPHALI
OF THE
ANCIENTS.

Of the Cynocephali *of the Ancients.*

IS not that I think there are any at present so mistaken, as to believe the *Cynocephali* to be a Race of *Men*, that I write this Essay : 'tis so notoriously known that they are *Monkeys*, or rather *Baboons*, that 'tis needless to go about to prove it, 'tis what even the *Ancients* themselves have sufficiently confessed. That which induces me to mention them, is to shew how fond the *Ancients* were of inventing *Fables* ; and *Ctesias*, who hath told us such fine Stories of the *Pygmies*, whom he makes to be *little Men*, tho' indeed they are only a sort of *Apes* ; when he comes to discourse of the *Cynocephali* , which are a sort of *Baboons*, and far less like *Men* than the *Pygmies* are, to perswade us that these likewise are a Race of *Men* ; he is obliged to exercise his Inventive Faculty with more force, to use much bolder strokes, and by roundly asserting so many incredible Things, to amuse our Imaginations, he hopes at least to give the Reader Entertainment, tho' he misses his Design of gaining our belief.

I will give you *Ctesias*'s own words, that you may see I do not abuse him,

him, as he hath done Mankind, in moſt of the *Natural Hiſto-y* that he hath left us ; for as (*a*) *Photius* informs us, *Cteſias* tells us

Ἐν τοῖς δὴ τοῖς ὄρεσι φησὶν ἀνθρώπυς βιοτδύειν, κυνὸς ἔχονίας κεφάλω. Ἐσθῆτας δὲ φοράειν ἐκ τῶν ἀγείων θη-είων· φωνὴν δὲ διαλέγον) ἐδεμίαν, ἀλλ' ὠρύον) ὥσπερ κύνες· κỳ ὕτω συνιᾶσιν αὐτῶ τ φωνήν. Ὀδόνlας δὲ μείζας ἔχειν κυνὸς, κỳ τὰς ὄνυχας ὁμοί-ως, κυνῶν, μακροτέρας δὲ κỳ ϛρογυ-λωτέρας. Οἰκεῖν δὲ ἐν τοῖς ὄρεσι, μέχρι τῦ Ἰνδῦ ποlαμῦ. Μέλανες δέ εἰσι κỳ δίκαιοι πάνυ, ὥσπερ κỳ οἱ ἄλλοι Ἰνδοὶ, οἷς [κỳ] ἐπιμίγνυν). Καὶ συνιᾶσι μὲν τὰ παρ' ἐκείνων λεγόμενα, αὐτοὶ δὲ ὐ δύναν) διαλέγεσθαι : ἀλλὰ τῇ ὀρυγῇ κỳ ταῖς χερσὶ, κỳ τοῖς δακτύλοις σημαίνεσιν, ὥσπερ οἱ κωφοὶ κỳ ἄλαλοι κỳαλῦ σι ͵ ἀπὸ τῶν Ἰνδῶν κỳ αλύϛριοι, ὅπες ἐϛιν ἑλληνιϛὶ κυνοκέφαλοι. Τὸ δὲ ἐθνΘ- ὅλον, ἕως δώδεκα μυειάδων.

Degere iiſdem hiſce in montibus homines memorat canino capite , qui ferarum pellibus veſtiantur. Sermo-ne hos nullo uti, canum tantùm more latratum edere, atque ita mutuò ſeſe intelligere. Dentes illis eſſe quàm canibus majores, & caninos ſimiliter ungues, ſed longiores , ac rotundio-res. Montes incolere ad Indum flu-vium uſque, & colore eſſe nigro, in-ſignéſque juſtitiæ cultores , ceterorum Indorum more, inter quos verſentur. Intelligere quoque quæ ab Indis di-cantur, licet ipſi loqui minimè poſſint, ut propterea latratu, manibus , atque digitis ſigna dent,quemadmodum ferè ſurdi ac muti ſolent. Vocari hos ab Indis Calyſtrios, quod Græci dice-rent κυνοκέφαλες, id eſt, Canicipites, [carnibus coſdem veſci crudis] totiúſ-que gentis capita numerari ad centum & viginti millia.

And a little afterwards he adds,

Ὅτι οἱ κυνοκέφαλοι οἰκοῦντες ἐν τοῖς ὄρεσιν, ὐκ ἐργάζονlαι : ἀπὸ θήρας δὲ ζῶ-σιν. Ὅτ' ἂν [δὲ] ἀποκτείνωσιν αὐτά, ἐπlῶσι πρὸς τ ἥλιον. Τρέφεσι δὲ κỳ προβατα πολλὰ, κỳ αἶγας, κỳ ὄϊς. Πί-νεσι δὲ γάλα κỳ ὀξύγαλα τῶν προβά-των. Ἐσθίεσι δὲ τ καρπὸν τῦ σιπα-χόρα, ἀφ' ὗ τὸ ἠλέκlρον· γλυκὺς γὰρ ὅϛι· κỳ ξυραίνοντες αὐτὸς, συνερίδας συρράπlεσιν, ὥσπερ ἐν τοῖς ἕλλησι τὴν ϛαφίδα. Οἱ δὲ κυνοκέφαλοι σχεδίαν ποιησάμενοι, κỳ ἐπlιθένlες, ἀπάγεσι φόρlον τέτα. Καὶ τῆς πορφύρας τὸ ἄν-θος καθαρὸν ποιήσανlες, κỳ τῦ ἠλέκlρε ξ' κỳ σ' τάλανlα τῦ ἐνιαυτῦ· κỳ ὕτω

Narrat inſuper hos Cynocephalos in montibus habitantes nullum exercere opificium ; ſed de venatione vivere, feráſque quas occiderint ad ſolem tor-rere. Magnam nihilominus pecoris co-piam alere, caprarúmque & ovium ; quarum quidem ovium lac atque oxy-gala pro potu illis ſit. Veſci tamen eti-am Sipachoræ fructu,è qua,uti dictum eſt, arbore ſuccinum emanat. dulcem enim illum eſſe. Hunc item illos fru-ctum arefactum in corbes conſtipare,ad eum modum quo uvas paſſa Græci. Eoſdem illos Cynocephalos ratem quoqz extemporariam conſtruere, qua impoſi-tum hujus fructûs onus, ut & purpuræ (ſed purgato prius ejus flore) itémque electri,ad ducenta & ſexaginta talen-

(a) *Photij Bibliothec. Cod. 72. de Indicis,* p. m. 149, &c.

τὸ

τὸ ῥοινίκινον ῥάπlει]τῷ φαρμάκῳ, ἕτερα
τοσαῦτα. Καὶ ἠλέκἰρε χίλια τάλαντα
ἀπάγειν κατ᾽ ἐνιαυτὸν τῷ Ἰνδῶν Βασιλεῖ. Καὶ ἕτερα]καὶ ἄγονlες πωλᾶσι τοῖς Ἰνδοῖς, πρὸς ἄρτες καὶ ἄλφιτα καὶ
ξύλινα ἱμάτια. Πωλᾶσι]καὶ ξίφη, οἷς
χρᾶν]ο πρὸς τὰ θηρίων ἄγρας, καὶ
τόξα καὶ ἀκόντια. Πάνυ γὸ καὶ δεινοί εἰσιν ἀκοντίζειν καὶ τοξεύειν Ἀπολύμηνοι
δ᾽ εἰσὶ, διὰ τὸ οἰκεῖν αὐτὰς ἄρεα ἄ-
βατα καὶ ὑψηλά. Δίδωσι]αὐτοῖς διὰ
πέμπ]ε ἔτες δᾶρα ὁ Βασιλεῦς, ά᾽ μὲν
μυριάδας τόξων, καὶ ἀκοντίων τοσαύτας,
πελ]ῶν]δώδεκα, καὶ ξίφη]πεντακισ-
μύρια. Τέτοις ταῖς κυνοκεφάλοις ἐκ
εἰσὶν οἰκίαι, ἀλλ᾽ ἐν σπηλαίοις διαιτῶν].
Θηρεύεσι]τὰ θηρία, τοξεύοντες καὶ
ἀκοντίζοντες, καὶ διώκοντες καταλαμβά-
νεσι ταχὺ γὸ τρέχεσι λέων]]αἱ
γυναῖκες αὐτῷ ἅπαξ τῇ μηνὸς, ὅταν
τὰ καταμήνια αὐτῆς ἔλθη ἄλλοτε δ᾽
ὄ. Οἱ]ἄνδρες ὀ λᾶτρῷ μὲν, τὰς δὲ
χεῖρας ἀπονίζον]. Ἐλαίῳ δὲ χρίον]αι
τρεῖς τῷ μηνὸς, τῷ ἀπὸ τὰ γάλακ]Θ-
γινομένῳ, καὶ ὀπλέζον] δέρμασι.
Τλὼ δὲ ἐσθῆτα ἔχεσιν ὀ δασεῖαν, ἀλλὰ
ψιλῶν τῷ μαθλημάτων. ὡς λεπ]ο-
τάτων, καὶ αὐτοὶ καὶ αἱ γυναῖκες αὐτῷ.
Οἱ δὲ πλυσιώτατοι αὐτῷ, λίνα φορέ-
σιν ὅτοι δὲ εἰσὶν ὀλίγοι. Κλῖναι δὲ αὐτοῖς ἐκ εἰσὶν, ἀλλὰ στιβάδας ποιᾶν].
Οὗτος δ᾽ αὐτῷ πλυσιώτατ]Θ- νομίζε]
ᾗ, ᾧ ἂν πλεῖςα πρόβατα ᾖ: καὶ ἡ
ἄλλη ἐσία, παραπλήσια. Οὐρὰν δὲ
ἔχεσι πάντες καὶ ἄνδρες καὶ γυναῖκες,
ὑπὲρ τῇ ἰσχύων· οἱάπερ κυνῶν μείζονα δὲ καὶ δασυτέραν. Καὶ μίσγον]
ταῖς γυναιξὶ τεἱραπόδηδὶ, ὥσπερ οἱ κύνες. ἄλλως δὲ μίγνυαι αὐτοῖς, ὀδὲν αἰ-
σχρὸν. Δίκαιοι δὲ εἰσι, καὶ μακροβιώ-
τατοι πάντων ἀνθρώπων· ζῶσι γὸ ἔτι
ρ᾽ καὶ ὁ᾽. ἔνιοι δὲ αὐτῷ καὶ τ᾽.

ta quotannis avehunt; *additis ta-
lentis totidem illius Pigmenti, quo
infectores puniceum colorem inducunt.
Electri praeterea mille talenta quotannis ad Indorum regem advehere. Im-
mò & alia plura devehere ad Indos
venalia, pro quibus viciffim panes,
farinam, & xylinas veftes accipiant.
Habere quoque enfes venales, quibus
ipfi ad venatum utuntur, cum arcubus
& jaculis. Peritiffimos enim effe jacu-
landi atque fagittandi: & praeterea
etiam, quod montes habitent altos atque inacceffos, bello infuperabiles. Regem ipfis ex munere quinto quoque anno praebere trecenta arcuum, totidém-
que jaculorum millia; jam peltarum
centum viginti, & gladiorum quin-
quaginta millia. Nullas item apud
hos effe domos. fed in antris degere.
In venatione jaculis potiffimum feras,
vel fagittis petere; eafdemque perfe-
quendo, quòd curfus velocitate prae-
ftent, etiam affequi. Horum uxores
femel duntaxat per menfem,cùm men-
ftrua patiuntur, lavare; aliàs nunquam. Neque viros unquam omnino
lavare, fed manus tantummodò ab-
luere. Oleo tamen ex lacte confecto ter
faltem menfibus fingulis ungi, & pellibus deinde abftergi.Vefte ad haec uti,
non villofa, fed è glabris maceratifq;
pellibus quàm tenuiffimis, ipfos aequè
atque uxores. Exceptis forte ditiffi-
mis inter eos, & iis quidem paucis,
qui lineos geftent amictus. Nec item
lectorum noviffe ufum eos, qui extem-
poraneos fibi toros exftruant. Hunc a-
pud eos ditiffimum haberi, qui plurimum habeat pecoris, ac reliquas opes
his propemodum effe fimiles. Caudam
infuper habere omnes, tam viros quàm*

mulieres, *fupra clunes, caninae, fimilem, nifi quòd major fit, & pilis den-
fior. Quadrupedes item hos, canum more, cum mulieribus congredi aliúm-
que congrediendi modum omnem pro turpi habere. Juftiffimos cofdem effe,
vitaeq; reliquos inter homines longiffimè. Vivere namq; ad centefimum ufq;&
feptuagefimum, nonnullofq; ad ducentiffimum quoq; annum.*   i. e.

' *i. e.* In thefe Mountains (faith *Ctefias*) live certain Men, who have
' Heads like Dogs, are cloathed with Skins of wild Beafts, fpeak no
' Language, but bark like Dogs, and thereby underftand one another.
' They have Teeth larger than Dogs; and Nails like Dogs, but longer
' and rounder. They dwell up in the Mountains, as far as the River
' *Indus*; they are black and very juft, as are the other *Indians* with whom
' they are mixt; and they underftand what is faid to them, tho' they
' cannot fpeak themfelves. But by their *Barking*, and their Hands and
' Fingers, they fignifie their Minds, as Deaf and Dumb Men do. They
' are called by the Indians, *Calyftrij*, which in *Greek* is *Cynocephali*. The
' whole Nation is an hundred and twenty Thoufand in number.

' Thefe *Cynocephali* that inhabit the Mountains, do not work, but live
' upon Hunting; and when they kill any wild Beaft, they roft it in the
' Sun. They breed a great many Sheep, Goats and Affes; and drink the
' Milk and Butter-milk of the Sheep. They eat likewife the Fruit of the
' *Sipachora* Tree, from whence comes *Ambar*, the Fruit whereof is fweet,
' which having dried, they put up in Baskets, as the *Greeks* do *Raifins*.
' Thefe *Cynocephali* having made a Boat, they load it with this Fruit, and
' with *Purpura*, the Flower being firft picked, and with *Ambar*, to the
' quantity of Two hundred and fixty Talents, which they every Year
' fhip off, and as much too of the Drug, with which the Dyers dye
' the Scarlet; and they carry every Year a Thoufand Talents of *Ambar*
' to the *King* of *India*, and they take with them other Commodities,
' which they fell to the *Indians*; for which they receive Bread, and
' Meal, and Cotton Garments. And the *Indians* fell them likewife Swords,
' which they ufe in taking the wild Beafts; and Bows and Darts, for they
' are very skilful Archers and Darters. They are invincible, becaufe
' they inhabit very high, and inacceffible Mountains. Every fifth year,
' the King beftows upon them Three hundred thoufand Bows, and as
' many Darts : Alfo an Hundred and twenty thoufand Shields, and Fifty
' thoufand Swords. They have no Houfes, but live in Caves. In hunt-
' ing the wild Beafts, they ufe their Bows and Darts, and purfuing
' them, they take them; for they run very fwift. The Women bathe
' only once a Month, when they have their *Catamenia*, otherwife not.
' The Men don't bathe, but only wafh their hands; but they anoint
' themfelves three times in a Month with Oyl made of Milk, and rub
' themfelves with Hides. The Cloaths both of the Men and Women are
' not hairy, but Skins macerated fmooth, and made very thin. The
' richeft of them wear Linnen, but thofe are but few. They have no
' Beds, but lye upon Straw or Leaves. He is efteemed the richeft a-
' mongft them, who hath moft Sheep, or fuch like Subftance. They have
' all, both Men and Women, Tails on their Rumps, like Dogs, but lar-
' ger and more hairy; and, like Dogs too, they lye with their Women
' on all four, and they think it unbecoming to do otherwife. They are
' juft

' juſt, and the longeſt lived of any Men, for they live an Hundred and
' ſeventy, and ſome of them Two hundred years.

Had not *Cteſias* made ſuch a Solemn Aſſeveration of the Truth of all
that he had wrote, that Apology that *Strabo* (*b*) makes for the *Poets*, might
excuſe him, Φαίνε?) γὴ (ſaith *Strabo*) εὐϑὺς ὅτι μύϑυς παρᾳπλέκυσιν ἑκόντες
ἐκ ἀγνοίᾳ τῇ ὄντων, ἀλλὰ πλάσει τῇ ἀδυνάτων, πρᾳ]είας ἡ τέρψεως κᾴριν·
i. e. *Statim enim apparet eos fabulas admiſcere, non ob verorum ignoratio-*
nem, ſed delectationis cauſa, monſtra & alia quæ eſſe non poſſunt , fingentes.
For our *Hiſtorian* had as good a Talent at Fiction as any of the *Poets*.
And tho' *Æſchylus*, as the ſame *Strabo* there tells us , firſt invented the
Story of the *Cynocephali*, or *Canicipites*, as likewiſe the *Pectoroculati*
and the *Unoculi*, as *Heſiod* and *Homer* did that of the *Pygmies* ; yet I
can't but think he hath as far out-done the Original in what he relates
of the *Cynocephali*, as he did in the Account he gives of the *Pygmies*.

Theſe *Cynocephali* by (*c*) *Ælian* are called ἄνθρωποι κυνοπρᴐσωποι, and he
gives this Relation of them, as I find him tranſlated by *Conrad. Geſner*,
who is more faithful in rendering him than *Pet. Gillius*, *Ultra Oaſin Æ-*
gypti, *ſolitudo maxima ad ſeptem dierum iter extenditur. Eam excipit Re-*
gio quam Cynoproſopi Homines incolunt, in via Æthiopiam verſus. Vivunt
illi Capraruni & Bubalidum venatu. Aſpectus iis niger, Caput & Dentes
Canis. Quod Animal, quum referant, non abſurda eorum (quamquam Hominum) *hoc in loco exiſtimanda eſt mentio. Nam & Sermonis uſu carent,*
& acuto quodam ſtridore ſonant. Barbam infra ſupraque os gerunt , Draco-
num quadam ſimilitudine. Manus eorum validis & acutiſſimis unguibus
armantur. Corpus omne hirſutum eſt , hoc etiam inſtar Canum. Sunt au-
tem perniciſſimi, & aquas Regionis norunt ; atque eam ob cauſam, difficiles
captu.

Now tho' *Ælian* calls them here *Men*, yet where he deſcribes them
before, even out of *Cteſias*, he plainly tells us they are not *Men*, but
only *Brutes*, becauſe they cannot ſpeak, but only bark. I will give you
(*d*) *Geſner*'s Tranſlation of this Paſſage : *In eodem Indiæ tractu, ubi Can-*
thari (†) *jam dicti, Cynocephali etiam reperiuntur : quibus à facie & Cor-*
poris ſpecie nomen inditum, cætera ferè humana habent : & veſtiti pellibus
ferarum ingrediuntur. Juſti ſunt, Hominum nemini moleſti aut injurij, non
Sermone ſed ululatu ſonant. Indorum tamen linguam intelligunt. Venatio-
ne Animalium ferorum vivunt, quæ ut ſunt celerrimi, facile conſecuti inte-
rimunt, & fruſtatim diviſa ad ſolem aſſunt. Capras etiam & oves alunt, ut

(b) *Strabo Geograph.* lib. 1. p. m. 29.   (c) *Ælian. Hiſt. de Animal.* lib. 10. cap. 25. p. m. 601.   (d) *Æ-*
lian. Hiſt. de Animal. lib. 4. cap. 46. p. m. 239.   (†) Theſe *Canthari* are that ſort of *Scarabæus* we
call a *Lady-Cow*, and I have formerly given a Figure of, in *Philoſoph. Tranſact.* N. 176. p. 1202. from
the Worm or Chryſalis of which , ſome the *Cochineel*, for dying Scarlet, of which there is a good
account here in .... of Cteſias.

ex lacte potu fruantur. Horum inter Animantes rationis expertes non teme-
rè mentionem feci, articulato enim, difcretoque & humano Sermone non u-
tuntur.

But 'twas for want of Education, it may be, and by their living wild
in the Woods, they loft their Learning and their Speech; for the Æ-
gyptians in the time of the *Ptolomies* took more care of them; and as the
fame Ælian relates, they taught them Letters, and to Dance, and to play
upon Mufick : Nor were they ungrateful to their Mafters; for they
beg'd a great deal of Money, which they carefully put up into a Bag, to
reward them for the pains they had taken with them. For thus, faith
(e) Ælian, as *Gefner* tranflates him; *Animalia Difciplinæ idonea hæc effe*
percepi. Regnantibus Ptolomeis *Cynocephalos Ægyptij literas, & faltare,*
& pulfare Citharam docebant. Tum verò unufquifpiam Cynocephalorum mer-
cedem, Domini nomine fic fcitè tanquam peritus aliquis Mendicus exi-
gebat. Et id quod dabatur in Marfupium , quod ferebat, appenfum, con-
gerebat.

I could not but take the more notice of this paffage in *Ælian*, becaufe
the *Cynocephali* are always reprefented to be of a fierce and untractable
Nature; which feems their particular Character : For faith *Ariftotle,* as I
have quoted him already (f), Καὶ οἱ κυνοκέφαλοι δὲ τὴν αὐτὴν ἔχϗσι μορφὴν
τοῖς πιθήκοις, πλὴν μείζονές τ' εἰσὶ, κ̀ ἰχυρότεροι, κ̀ τὰ πρόσωπα ἔχοντες κυνοειδέςερα. Ἔτι δὲ ἀγριώτερά τε τὰ ἤθη, κ̀ τὰς ὀδόντας ἔχϗσι κυνοειδεςέρϗς κ̀
ἰχυροτέρϗς. i. e. The Cynocephali *are of the fame fhape with Monkeys , but*
they are bigger and ftronger, and they have a Face more like a Dog's, and are
of a fiercer Nature, and have Teeth more like a Dog's, and ftronger. And
fo (g) *Pliny, Efferacior Cynocephalis Natura ficut Satyris :* And (h) *Solinus,*
Cynocephali & ipfi funt è numero fimiarum, in Æthiopiæ parte frequentiffimi :
violenti ad faltum, feri morfu, nunquam ita manfueti, ut non fint magis ra-
bidi. And (i) *Diodorus Siculus* defcribes them after the fame manner :
Οἱ δ᾽ ὀνομαζόμενοι κυνοκέφαλοι τοῖς μὲν σώμασιν ἀνθρώποις δυσιδέσι παρεμ-
φερεῖς εἰσὶ, ταῖς δὲ φωναῖς μυγμὲς ἀνθρωπίνϗς πρείεν)· ἀγριώτατα δὲ ταῦτα τὰ
ζῶα, κ̀ παντελῶς ἀπιθάσσδλα καθεςῶτα, τὴν ἀπὸ τῶ̀ ὀφρύων πρόσοψιν σκυθ-
ρωτέραν ἔχει ταῖς δὲ θηλείαις ἰδιώτατόν συμβαίνει, τὸ τὴν μήτραν ἐκτὸς τῶ σώ-
ματος φέρειν ἅπαντα τ̀ χρόνον. i. e. *Qui* Cynocephali *(à Canino Capite) di-*
cuntur, Corporis afpectum Hominum deformium inftar habent, quorum vocem
muffitatione tantùm exprimunt. Apprimè ferox eft hoc Animal, nec ullo ci-
curationem pacto admittit, & vultum à fuperciliis aufterioribus præfert. Sin-
gulare quiddam femellis accidit ; quod vulvam perpetuò extra Corpus projectam
habent. And *Agatharcides* in (k) *Photius* gives juft the fame defcription
of them : Ὁ δὲ κυνοκέφαλΘ (faith he) τὸ μὲν σῶμα ἀνθρώπϗ δυσιδϗς ὑπ-

---

(e) *Ælian. Hift. de Anim. lib. 6. cap. 10. p. 331.* (f) *Vid. pag. 3. & 7.* of the Anatomy. (h) *Plinij*
Hift. Nat. lib. 8. cap. 54. p. 243. (h) *Solini Polyhiftor. cap. 27. p. m. 39.* (i) *Diodori Siculi Bi-*
bliath. Hiftor. lib. 3. p. m. 158. (k) *Photij Bibliath. Cap. 38. Cod. 250. p. m. 1364.*

G

χοράψει

γεφρει, τὸ πρόσωπεν δὲ κυνὸς᾽ φωνὴν δὲ ἀφίησι μυγμῷ παραπλησίαν᾽ ἄγριον
δὲ ὑπερβολῇ ἢ τελέως ἀτιθάσευτον, ἢ τὴν ὄψιν ἐμφαίνον ἀπό τε τῶν ὀφρύων καὶ
τῶν ὀμμάτων αὐσηράν. Περὶ μὲν ᾗ ἄρρενα ταῦτα. Τῷ δὲ ϑήλει περόσκει
καὶ τὸ τὴν μήτεραν ἔξω τὸ σώματᾼ φορεῖν, καὶ οὕτω διαγίνεϑαι πάντα ᾗ βίον.
i. e. *In Cynocephalo Hominis Corpus, specie turpis, adumbratur. Canina ei
facies, vocem stridori Muris non dissimilem exprimit. Sed immodicè ferum
est Animal, nec ullo modo cicuratur: vultumque à superciliis & oculis auste-
rum prodit. Ita Mas comparatus est. Fæmineo generi hoc est proprium, ut
uterum extra Corpus gestet, eoque habitu totam exigat vitam.*

    Salmasius (1) remarks that *Agatharcides* borrowed this Passage, as he
hath some others likewise, from *Diodorus Siculus.* But that these Relations of *Ctesias* that are so extravagant and wild, should be copied from
him, by so many and noted Authors too, seems somewhat strange. Yet
we find *Ælian, Pliny, Solinus,* and a great many others have done it ;
tho' they have added by it little Credit to their Histories, and no doubt
much lessened their own Reputation by transcribing the Errors of their
Predecessors. In the *History* therefore of *Nature* we must not depend
upon the Authority of the ·Number of those that only transcribe
the same thing, without duly examining the Matter themselves : For the
Authority here wholly depends on the veracity of the first Relator : And
if what *Ctesias* saith is false, tho' never so many say the same thing from
him, they must all be in the wrong. Especially in transcribing the Ancients, and believing their Reports, we ought to be very cautious, since 'twas
a common Practice amongst them to disguise and conceal the Truths they
would deliver, in *Ænigmatical* and *Mythological* Representations. Many
times there is something of Truth contained in their Relations, but 'tis
under such Vails, that you will not discover it, till you have taken them
off. And tho' there are no such *Men,* as *Ctesias's Cynocephali,* and *Pygmies;*
yet there are *Apes,* and *Monkeys,* and *Baboons,* that afforded him a ground
for his Invention.

    Now what sort of *Monkey* these *Cynocephali* were, I shall not at present
enquire ; that they are of the *Monkey-kind* is evident, because they have
Tails : and *Aristotle* tells us, that they are bigger and stronger, and therefore I make them of the *Baboon-kind.* But not having seen any of them
my self, I shall refer my Reader to the Authors who have wrote about
them. 'Tis sufficient to my present purpose that they are a sort of *Monkeys,* and not *Men,* as formerly represented.

    (1) *Salmasij Exercitat. Plinian. Cap.27.p.267.*

A

A
Philological Essay
Concerning the
SATYRS
OF THE
ANCIENTS.

Of the SATYRS of the Ancients.

TULPIUS and *Bontius* indeed think the *Orang-Outang* to be the *Satyr* of the Ancients ; but if we enquire into their History, and examine what Opinion the Ancients had concerning them, we shall find it no less involved in *Fables,* than that of the *Pygmies* ; and upon this account several of our Learned Men of late, have wholly denied them, and look upon all the Stories concerning them to be only a Fiction of the *Poets* and *Painters,* and that there were never any such Beings in Nature. The Learned (*a*) *Casaubon* is clearly of this Mind, *Quicquid de Satyris legimus* (saith he) *ex Poetarum Pictorumque fingendi Licentia Originem ducere. Nihil hujus reverà in Rerum Naturâ existere.* So (*b*) *Isaac Vossius* speaking of the *Ægipanes* tells us, *Sanè neque in formâ hujus monstri conveniunt, si tamen monstris accensenda sunt ea, quæ sunt mera Græculorum Commenta.* And the Learned (*c*) *Bochartus* saith, *Absit interim ut ex his locis Quisquam colligat, ullos aut jam exstare, vel unquam extitisse in Rerum Naturâ Satyros.* However, I do not doubt but to make it plainly appear, that there were

---

(a) *Casaubon de Poës Satyricâ,* lib. 1. cap. 2.   (b) *I. Vossii Comment. ad Pompon. Melam.* lib. 1. c. 8. p. m. 45. (c) *Bocharti Hierozoic. seu de Animal. Sacræ Scripturæ.* part. post. lib. 6. cap. 7. p. 829.

           such

such Animals in *Africa* which the Ancients called *Satyrs*. And tho' they sometimes called them *Men*, and for the most part worshipped them as *Gods*, yet I shall shew, that they were only a sort of *Monkeys*, and likewise Evince, that the *Orang-Outang* was not this sort of *Monkey* or *Satyr* of the Ancients.

Having proposed these as the Heads of my ensuing Discourse, it will not be expected of me to give an Account of all that has been said on this Argument. I shall rather apply my self to make out what I have here asserted. And tho' on this occasion, it may be, the *Poets* have *Enigmatically* represented some Nobler Secrets of *Philosophy*, by what they relate under the *Fables* they have made of these *Satyrs*, the *Fauni*, the *Nymphæ, Pan, Ægipan, Sylvanus, Silenus*, or any other Name they have given of this sort of *Animal*; yet I think my self no farther concerned at present, than to shew what might give the first rise to and occasion of these Inventions : or rather to prove that the *Satyrs* were neither *Men*, nor *Demi-gods*, nor *Dæmons*; but *Monkeys* or *Baboons*, that in *Africa* were worshipped as the *Gods* of the Country; and being so, might give the *Poets* the Subject of the Stories which they have forged about them.

The *Satyrs* therefore are generally represented like *Men* in the upper Parts, but with Horns on their Heads; and in their lower Parts or Legs like *Goats :* hence they are called *Capripedes*, or Αἰγίποδἐς ἄνδρες, as *Herodotus* expresses it. And *Pliny* (as I shall shew) where he describes them as *Brutes*; and saith, they are sometimes *Quadrupeds*, sometimes *Bipeds*, yet tells us, they are *Humanâ Effigie*. *Diodorus Siculus* (*d*) informs us, that when *Osiris* went into *Æthiopia*, ἀχθῆναι λέγεσι πρὸς αὐτὸν τὸ τῶ Σατύρων γένΘ., ὅς φασὶν ὅπὶ τῆς ὀσφύΘ. ἔχειν κόμας,&c. i. e. *Dum in Æthiopia versatur (Osiris) Gens Satyrorum ei adducitur , quas pilos in lumbis (Osphye) habere ferunt. Risûs enim amator erat Osiris & Musicæ Choreisq; gaudebat*, &c. *Satyri igitur quia ad tripudia, & decantationem Carminum, omnemque hilaritatem & lusum apti erant, in partem Militiæ venerunt.* He makes them likewise the Companions of *Bacchus*, and for the same reason (*e*), Τὰς ἢ Σατύρες ταῖς πρὸς γέλωΐα συνεργέσαις ὑπηλησίοσιν χρωμένας, παρασκευάζειν τῷ Διονύσῳ τὸν εὐδαίμονα ἠ κεχαρισμένον βίον· i. e. *Ita Satyri ludicris & ad risum compositis gestibus & actionibus, vitam Dionyso beatam, Gratiisque delibatam, reddunt.* And they are always represented as Jocose and Sportful, but Scurrilous and Lascivious; and wonderful Things they relate of their Revellings by Night, their Dancing, Musick, and their wanton Frolicks. For thus *Pliny* (*f*) describing the Parts about the Mountain *Atlas* in *Africa*, informs us, *Incolarum neminem interdiu cerni : silere omnia, non alio quam solitudinum horrore : subire tacitam*

(*d*) *Diodorus Siculus Biblioth:ec. Hist.* lib. 1. p. m. 16.   (*e*) *Diodorus Siculus ibid.* lib. 4. p. m. 213.
(*f*) *Plinij Hist. Nat.* lib. 5. cap. 1. p. m. 523.

Re-

Religionem animos propriùs accedentium, præterque horrorem elati (sc. Montis) super nubila, atque in viciniam Lunaris circuli. Eundem noctibus micare crebris ignibus, Ægipanum, Satyrorumque lascivia impleri, Tibiarum ac Fistulæ Cantu, & Cymbalorum Sonitu strepere.* And then adds, *Hæc celebrati Authores prodidere.* And so (g) Pomponius Mela, *Ultra hunc sinum Mons altus (ut Græci vocant) Θεῶν ὄχημα, perpetuis ignibus flagrat : ultra montem viret Collis longo tractu, longis littoribus obductus , unde visuntur patentes magis Campi, quàm ut prospici possint, Panum, Satyrorumque. Hinc opinio ea fidem cepit, quòd cum in his nihil culti sit, nullæ habitantium Sedes, nulla Vestigia, solitudo in diem Vasta, & silentium Vastius, nocte crebri ignes micant, & veluti Castra late jacentia ostenduntur, Crepant Cymbala & Tympana, audiunturque Tibiæ Sonantes majus humanis.* Where we may observe that what *Pliny* calls *Ægipanes,* *Mela* calls here *Panes.* And the *Satyrs* being commonly called *Fauni,* I can't but think, that the idle Stories we have about the *Fairies,* must come from hence: For they likewise have their Revellings, Dancing, and Musick by Night. And as even to this day, to fright Children, they tell them Stories of *Fairies* and *Hobgoblins,* so the Ancients did use to call any great sudden Fear, as we do now, a *Panick Fear,* from this *Pan.* For as (h) *Pausanias* tells us, Ἐν ᾗ τῇ νυκτὶ φό&c. σφίαν ὑπῆλθε Πανικές. Τὰ γὸ ἀπὸ αἰτίας ἐδεμιᾶς δείματα ἐκ τότε φασὶ γίνεθαι· i. e. *Eâ nocte Panicus illos incessit terror. Terrores enim nulla ex causa Ortos ab eo (sc.* Pane*) immitti aiunt.* And so (i) *Euripides* :

Κεσνὶς Πανὸς τρομερᾷ μάςιγι φοβῇ.

Saturnij (Senis*) Panis tremendo flagello (* Ictus*) trepidas..*

And so (k) *Dionysius Halicarnasseus* speaking of the *Faunus,* says, Τέτῳ γὸ ἀνατιθέασι τῷ δαίμονι Ῥωμαῖοι τὰ Πανικὰ, ᾗ, ὅσα φάσμαλα ἃ ὅτε ἀλλοίας ἴχονλα μοςφὰς, εἰς ὄψιν ἀνθςώπων ἔρχον), δείμαλα φέρονλα· i. e. *Huic enim Romani Panicos terrores adscribunt, & quæcunque alia Spectra, quæ varias induentia formas in Hominum conspectum veniunt, & Metum ipsis incutiunt..* And (l) *Ovid* :

————*Faunique bicornes Numine contactas attonuere suo.*

How jolly therefore soever and merry the *Satyrs* may be by night amongst themselves, with their Dancing and Musick : yet they have been frightful to Men formerly, as the Stories of the *Fairies* and *Hobgoblins* are (as I said before) to Children now ; and indeed, the telling Children Stories of this

---

(g) *Pomp. Mela de situ Orbis,* lib. 3. cap. 9. p. m. 63.   (h) *Pausanias in Phocicis.*   (i) *Euripides in Rhesò.*   (k) *Dionysij Halicarnass.* lib. 5. cap. 3.   (l) *Ovid in Phædra.*

kind,

kind, is a very mischievous Custom; for they are thereby impressed with such Fears, as perhaps they cannot conquer all their Life time.　But the Account that (*m*) *Phurnutus* gives of these *Panick Fears*, I think is natural; for he tells us, "Ἔςι δ῾ τὸ πανικὰς λέγεδαὶ παραχὰς τὰς ἀφνιδίες, κ᾽ ἀλόγες, ἔτω γὰρ πως κ᾽ αἱ ἀγέλαι, κ᾽ τὰ ἀἴπόλια ποιεῖται, ψόφε τινὸς ἐξ ὕλης, ἤ τῶ᾽ ὑποκάντ᾽εων κ᾽ φαραγᾳ῾ωδ᾽ων τόπων ἀκέσαντα· i. e. *Nihil prohibet quin etiam Panicos tumultus dicamus, qui subito & sine ratione certa exoriuntur : sic enim interdum armenta & greges terrentur, dum sonus quidam subitus è Sylva, aut ex Antris aut ex Terræ voraginibus affertur.*

Now *Lucretius* thinks that all this Musick of Pipes, Flutes, Cymbals and Drums, that is said to be made by the Jollity and Revellings of the *Satyrs, Fauni, Panes,* &c. in this dreadful Mountain by Night, is meer Romance and Fiction; and that 'tis nothing but the *Eccboing* of the whistling boisterous Winds amongst those hideous Rocks : For speaking of *Eccho's*, he tells us (*n*),

> *Sex etiam, aut septem loca vidi reddere voces*
> *Unam cum jaceres : ita colles collibus ipsis*
> *Verba repulsantes iterabant dicta referre.*
> *Hæc loca Capripedes Satyros, Nymphasque tenere*
> *Finitimi fingunt, & Faunos esse loquuntur ;*
> *Quorum noctivago strepitu, Ludoque jocanti*
> *Adfirmant vulgo taciturna silentia rumpi,*
> *Chordarumque Sonos fieri, dulceisque querelas,*
> *Tibia quas fundit digitis pulsata canentum :*
> *Et genus Agricolùm late sentiscere, cum Pan*
> *Pinea semiferi Capitis velamina quassans,*
> *Unco sæpe labro calamos percurrit hiantes,*
> *Fistula Silvestrem ne cesset fundere Musam.*
> *Cætera de genere hoc monstra, ac Portenta loquuntur,*
> *Ne loca deserta ab Divis quoque fortè putentur*
> *Sola tenere : ideo jactant miracula dictis*
> *Aut aliqua ratione alia dicuntur, ut omne*
> *Humanum Genus est avidum nimis auricularum.*

Which the Ingenious Mr. *Creech* hath thus rendered :

> —— And I my self have known
> Some Rocks and Hills return *six* words for *one :*
> The *dancing* words from Hill to Hill rebound,
> They all *receive*, and all *restore* the sound.
> The *Vulgar*, and the Neighbours think, and tell,
> That there the *Nymphs*, and *Fauns*, and *Satyrs* dwell ;

(*m*) *Phurnutus de Natura Deorum Cap. de Pane,* p. m. 70.　(*n*) *T. Lucretij de Rerum Natura,* lib. 4. vers. 581.

And

And that *their* wanton fport, *their* loud delight
Breaks thro' the *quiet* filence of the Night :
Their *Mufick's* fofteft Ayrs fill all the Plains,
And mighty *Pan* delights the liftning Swains ;
The *Goat-fac'd Pan,* whilft Flocks fecurely feed,
With *long-hung lip* he blows his Oaten Reed ;
The horn'd, the half-beaft God, when brisk and gay
With Pine-leaves crown'd, provokes the Swains to play,
Ten thoufand fuch *Romants* the Vulgar tell,
Perhaps leaft Men fhould think the Gods will dwell
In *Towns* alone, and fcorn their *Plains* and *Cell*
Or fomewhat ; for Man *credulous* and *vain*
Delights to *hear* ftrange things, delights *to feign.*

Lucretius here attributes the Invention of thefe Fables to the fuperftitious Notions Men had of *Deities,* and the Itching Ears Mankind generally hath for hearing Novelties and Wonders ; and no doubt,the fatisfying this Humour put the *Ancients* upon inventing moft of thefe Stories. But we may take notice that *Lucretius* places together the *Satyrs,* the *Nymphs,* the *Fauni* and *Pan* ; and generally I obferve, where mention is made of them, feveral are joyned together : As *(o) Ovid,*

> *Illum Ruricolæ, Sylvarum Numina, Fauni*
> *Et Satyri fratres, & tunc quoque clarus Olympus*
> *Et Nymphæ flerunt.*

The *Fauni* therefore and *Satyrs* I find are near akin.   And *(p) Ovid* in another place faith,

> *Quid non & Satyri Saltatibus apta juventus*
> *Fecere, & Pinu præcincti Cornua Panes.*
> *Silvanufque fuis femper juvenilior annis.*

And elfewhere he tells us *(q),*
> *Panes & in Venerem Satyrorum prona juventus.*

The *Satyrs* therefore and *Fauni* feem to be young ones, and the elder, the *Panes* and *Silvani,* according to that of *(r) Virgil,*

> ———*Deos qui novit agreftes*
> *Panaque, Silvanumque Senem Nymphafque Sorores.*

And *(s) Plutarch* tells us that what the *Greeks* called *Ægipan,*the *Romans* called *Silvanus.*   And *(t) Paufanias* exprefly tells us, that when the *Satyrs*

(o) *Ovid. Metamorph.* lib. 6. verf. 392.   (p) *Ovid. Metamorph.* lib. 14. verf. 637.   (q)*Ovid.* l.1.
(r) *Virgil.Georg.*l.2.verf.494.   (s) *Plutarch.in Parallelis.*   (t) *Paufan.in Attic.*p.m.21.

grow old, they are called *Sileni* : Τὸς γὸ ἐλικίᾳ τῶν Σατύρων προβηκοντας ὀνομάζεσι Σειληνὸς.   And by *Virgil*'s Expreſſion *Nymphaſque Sorores*, 'tis very evident, that the *Nymphs* likewiſe were of this Family, and nearly related.   *Ovid* (*u*) joyns them together.

> *Sunt mihi Semidei, ſunt Ruſtica Numina Nymphæ,*
> *Faunique, Satyrique, & monticulæ Silvani.*

Now what difference there is amongſt all theſe, unleſs as to their *Age* and *Sex*, I will not undertake at preſent to determine.   The *Poets* and the *Painters* of old, if we nicely enquire into them, have been pleaſed, as their fancy govern'd them, to make, or not make a diſtinction between them.   Thoſe that have a mind to ſatisfie their Curioſity farther in this Matter, may conſult *Salmaſius*, *Bochart*, *Gerard*, and *Iſaac Voſſius* , and ſeveral others, who have largely wrote about them.   I am apt to think that *Pan*, *Ægipan*, *Silvanus* and *Silenus* , were all the ſame ; as were the *Satyri* and the *Fauni* ; only theſe were younger than the former ; and the *Nymphs* were the *Females* of the Kind. But 'tis ſufficient to my buſineſs, if I make it appear, notwithſtanding all this, that the *Satyrs* were not *Men*, nor *Demi-Gods*, nor *Dæmons*, but only Brutes of the *Monkey-kind* ; which is plain enough even from the Ancients , who have invented ſo many Fables about them.

For (*w*) *Herodotus* tells us, and he is apt enough oftetimes to be over-credulous,ἔμοὶ μὲν ὁ πρὸ͂α λέγοντες,οἰκέειν τὰ ἔρεα ἀγριπόδες ἄνδρες· for they are neither Men, nor have they ſuch Feet. *Satyri de hominibus nihil aliud præferunt quàm figuram*, ſaith (*x*) *Solinus*. *Satyrus præter Effigiem nihil humani*, ſaith (*y*) *Mela*. *Pliny* gives us a larger deſcription of them; *Sunt & Satyri* (ſaith (*z*) he) *ſubſolanis Indorum montibus (Catharcludorum dicitur Regio) pernicioſiſſimum Animal : Cùm Quadrupedes tum rectè incedentes, humanâ effigie, propter velocitatem, niſi Senes aut ægri, non capiuntur. Choromandarum Gentem vocat Tauron, Silveſtrem, ſine voce, ſtridoris horrendi, hirtis Corporibus, oculis glaucis, dentibus caninis.*   You may here perceive they have ſomething of the ſhape of Men, but can't ſpeak, they are hairy, they go ſometimes upon all four, ſometimes erect, they have Dogs Teeth, they are wild miſchievous Animals. But *Ælian* is a little more expreſs : *Finitimos* Indiæ *montes* (ſaith (*a*) he) *tranſmittenti, ad intimum latus denſiſſimas convalles videri aiunt, &* Corudam *locum nominari : ubi Beſtiæ Satyrorum ſimilitudinem formamque gerentes, & toto Corpore hirſutæ, verſantur : atque Equina Cauda præditæ dicuntur. Eæ quum non à venatoribus agitantur , in opacis & ſpiſſis Sylvis ſolent ex frondibus (& fructibus) vivere. Quùm autem Venantium ſtrepitum ſentiunt, & Ca-*

(u) *Ovid. Metamorph.* lib. 1. verſ. 193.   (w) *Herodot.* in *Melpomene*, p. m. 229.   (x) *Solinus Polyhiſt.* cap. 34.   (y) *Pomp. Mela de ſitu Orbis*, lib. 1. cap. 8. p. 11.   (z) *Plinij Hiſt. Nat.* lib. 7. cap. 2. (a) *Ælian. Hiſt. Animal.* lib. 16. cap. 21.

mmm

num latratus exaudiunt, in Montium vertices incredibili celeritate excur-
runt : nam per montes iter conficere assuetæ sunt. Contra eos qui se inse-
quuntur pugnant , de summis montibus saxa devolventes, quorum impetu
sæpe multi deprehensi pereunt. Itaque difficillimè capiuntur : Et ex iis non-
nullæ, sed ægerrimè tandem, aut ægrotantes nimirum, aut gravidæ compre-
henduntur. Illæ quidem propter morbum ; hæ verò ob gravitatem. Captæ
antem ad Prasios deferuntur. Ælian here tells us that they have Tails like
Horses, therefore they must be of the *Monkey* or *Baboon* kind. And
Pausanias, who made it his Business to enquire more particularly about
them, informs us they have such Tails, but can't speak, but are very Lascivious and Lustful, as they are observed to be to this day. I will give
you *Pausanias's* words ; Περὶ ἢ Σατύρων (saith (*b*) he) οἵτινες εἰσὶν, ἐτέρις
πλέον ἐθέλων ὑπίσαθαι, πολλοῖς αὐτῶ τέτων ἕνεκα ἐς λόγοις ἦλθον. Ἔφη ἢ
Εὔφημ₡. Κάρ ἀνὴρ πλέων ἐς Ἰταλίαν ἁμαρτεῖν ὑπ' ἀνέμων τῶ πλῦ, ἡ ἐς τ'ω
ἔξω θάλασσαν,ἐς ἢν ἔκέτι πλέυσιν ἐξεννεχθῆναι Νήσοις ἢ εἶναι μὲν ἐρήμες πολλάς,
ἐν ἢ ταύταις εἰκεῖν ἄνδρας ἀγρίες· ἄλλαις ἢ ἐκ ἐθέλειν νήτοις προστέχειν τὰς
ναύτας, οἷα πρότερόν τι προσχόντας,ἡ τῆ ἀνοίων ἐκ ἀπείρως ἔχοντας. Βιασθῆναι
δ'ὗν ἡ τότε. Ταύτας καλεῖσθαι μὲν ὑπὸ τῆ ναυτῶ Σατυρίδας, ἡ ἢ τῆς ἀνοι-
κέντας ἡ πυρρὰς, ἡ ἵππων ὁ πολὺ μείες ἔχειν ὑπὶ ταῖς ἰσχίοις ἐρᾶς. Τέτυς ὡς
ἤσονο κατασδραμύντας ὑπὶ τὼ ναῦ. φωνὼ μὲν ἐδεμίαν ἱέναι, ταῖς ἢ γυναιξὶν
ὑπιχειρεῖν ταῖς ἐν τῆ νηί. Τέλ₡. ἢ δείσαντας τὰς ναύτας, βάρβαρον γυναῖκα
ἐκβαλεῖν ἐς τὼ νῆτον. Ἐς ταύτην ὗν ὑβρίζειν τὰς Σατύρος, ἢ μόνον ἡ καθῆκεν-
κεν, ἀλλὰ τὶ τὸ πᾶν ὁμοίως σῶμα. Which (*c*) *Conrad. Gesner.* I find hath
thus translated ; *Cæterum de Satyris, quinam sint, cùm plura quàm alij scire*
laborem, cum multis ea de re sum collocutus : Dixit autem Euphemus Car, se
quùm in Italiam navigaret cursu esse excussum vi ventorum, & ad mare exti-
mum, quod navigari non item soleat portatum. Insulas autem ibi multas esse
ac desertas, & viris agrestibus incoli. Ad alias verò aiebat nautas deflectere
recusasse, quòd antea quoque eò appulsi, Incolarum Inhumanitatem essent ex-
perti. Tempestatis denique violentia eò pervenisse. Insulas eas à Nautis vo-
cari Satyrias. Incolas inesse rubicundos , & caudas imo dorso habere, Equi-
nis non multò minores. Hos, ubi senserant, ad navigium accurrisse, nullam-
que vocem edidisse , sed mulieribus Navi unà advectis manus injecisse. Nau-
tas verò timore correptos, Barbaram Mulierem in Insulam tandem projecisse.
Eam Satyros, non solum qua parte consuetudo permittat, verùm etiam toto
corpore libidinosè violasse, referebat.

It appears therefore plainly that the *Satyrs* have Tails. But that there
might not the least Scruple remain what sort of Animals these *Satyrs* were,
I shall produce a Passage out of (*d*) *Philostorgius* which is very express,
and comes fully up to our Business : For he tells us , Ἔςι ἢ ἡ τῦτο (sc.
Satyrus) πιθηκ₡., ἐρυθρὸν τὸ πρόσωπον , ἡ γοργὸς τὴν κίνησιν, ἡ ἐσρὰ ἔχων.
i. e. *That a Satyr is a sort of Ape with a red face, swift of motion, and ha-*

(b) *Pausanias in Atticis,* p. m. 21. (c) *Gesner. de Animal.* p. 865. (d) *Philostorgij Hist. Eccle-*
siastic. lib. 3. cap. 11. p. 41.

H *ving*

ving a Tail. Where you may obferve that *Philoſtorgius* and *Pauſanias* both agree, that they have a red Face, which may be fome mark, by which to know them again. And (e) *Galen* hath given us another, *viz.* that their *Roſtrum* or Chin is longer than an *Apes*, but not ſo long as that of the *Cynocephalus*, as appears in that Paſſage I have already quoted (f), viz. *That a Man in proportion to his Body hath the ſhorteſt Chin of any* Animal ; *next to a* Man, *an* Ape ; *then the* Lynx *and* Satyrs ; *and after theſe the* Cynocephali. Now none of theſe Marks agree to the *Orang-Outang* ; for it had no *Tail*, it had not a *red Face*, and his *Chin* was *ſhorter* than any other ſort of *Apes*. So that *Bontius* was miſtaken in calling it a *Satyr*. And *Tulpius* was too haſty in laying down this Conclufion, *In ſumma* (ſaith (g) he) *vel Nullus eſt in Rerum Naturâ Satyrus : aut ſi quis eſt, erit proculdubio illud Animal, quod in Tabellâ hic à nobis depiⱷum.* Had *Tulpius* a mind to have made his *Orang-Outang* a *Satyr*, he ſhould not have compared him to a Courtier, nor inſtanced in ſuch Niceties as he obſerves, of his drinking, and going to bed : For, *Efferatior Cynocephalis Natura,ſicut Satyris,* ſaith (h) *Pliny.*And in another place he tells us,*Satyris præter figuram nihil moris humani* (i). But the *Orang-Outang* had very tender Paſſions,and was very gentle and loving. Another very remarkable difference that I find between the *Satyrs* and the *Orang-Outang*, is, that the *Satyrs* have Pouches in their Chops as *Monkeys* have ; but the *Orang-Outang*, as I have ſhewn in the Anatomy, had none. *Condit* (ſaith (k) *Pliny*) *in ▪Theſauros Maxillarum Cibum Sphingiorum & Satyrorum genus : mox inde ſenſim ad mandendum, manibus expromit : & quod formicis in annum ſolenne eſt, his in dies vel horas.* The *Orang-Outang* therefore cannot be the *Satyrs* of the Ancients, as *Tulpius*, and *Bontius*, and *Dapper* imagined.

By what has been ſaid,I think it fully appears that there were ſuch *Animals* as the Ancients called *Satyrs* ; and that they were a ſort of *Monkeys* or *Apes* with Tails : And this Account that I have given of them, will very well make out thoſe Texts in *Iſaiah* ; as *Chapter* 13. *verſe* 21. *But wild Beaſts of the Deſart ſhall lye there, and their Houſes ſhall be full of doleful Creatures, and Owls ſhall dwell there, and Satyrs ſhall dance there.* And *Chapter* 34. *verſ.* 14. *The wild Beaſts of the Deſart ſhall alſo meet with the wild Beaſts of the Iſland ; and the Satyr ſhall cry to his Fellow* ; *the Schrich-Owl alſo ſhall reſt there, and find for her ſelf a place of reſt.* For ſince the Text calls them wild Beaſts, I ſee no reaſon why we ſhould fancy the *Satyrs* here to be *Dæmons*, as the Learned *Bochartus* and others ſeem to do. I agree with *Bochart*, that what is told us in the Life of St. *Paul* the Hermite by St. *Jerome* , and in that of St. *Anthony* by St. *Anaſtaſius* of a *Satyr* meeting St. *Anthony* in the Deſart,and diſcourſing with him, may be

(e) *Galen. Adminiſtr. Anat.* lib. 4. cap. 3. p. m. 94.   (f) *Vid. Anat.* of the *Orang-Outang* , pag. 94.
(g) *Nk. Tulpij Obſerv. Med.* lib. 4. cap. 56. p. m. 274.   (h) *Pliny Nat. Hiſt.* lib. 8. cap. 54. p. 243.
(i) *Plinj ibid.* lib. 5. cap. 8. p. m. 549.   (k) *Plinij Nat. Hiſt.* lib. 10. cap. 72. p. m. 466.

fabulous

fabulous or a Delusion. *Non assentior* (saith *(l) Bochartus) narrationi Magni Scriptoris, in qua Satyrus introducitur Antonium in Eremo rogans, ut pro se communem Deum deprecetur, tanquam Salutis in Christo particeps futurus. Non alios servat Christus, quàm quos assumpsit. At non assumpsit Angelos, multò minùs Dæmones ant Satyros qui nusquam sunt, sed semen Abrahæ.* And tho' St. *Jerome*, to confirm this Relation, adds, That in *Constantine's* time one of these Monsters was seen alive at *Alexandria* in *Ægypt*, and after it's Death, it's Carcals was embalmed and sent to *Antioch* for the Emperor to see it; Yet I shall plainly prove that this *Satyr* was nothing else but that sort of *Monkey* I am now discoursing about.

This Story I find often mentioned; but *(m) Philostorgius* gives us the most particular Account of it, and therefore I shall insert his own words; Καὶ ἄλλαις πολλῶν ζωῶν εἰδέαις; τῆς πιθηκείας μορφῆς ὑπομιμουμένης· ἢ δῆλον ταῦτα θεῖν, πολλῶν εἰς ἡμᾶς κομιζομένων, οἵ. δὴ ἢ ὁ Πὰν ὑπαληθεῖς ὑπάρχει, ὃς τὼ κεφαλὼ αἰγοπρόσωπός θτι, ἢ αἰγίκερως, ἢ ἐκ λαγῶνων τὰ κάτω αἰγοσκελής, τὼ δὴ κοιλίαν ἢ τὸ στέρνον ἢ τὰς χεῖρας καθαρὸς πίθηκ@, ὃν ἢ ὁ τῶν Ἰνδῶν Βασιλεὺς Κωνσαντίῳ ἀπέςαλκει. Τὅτο δὴ τὸ ζῶν ἔζη μὲν φερόμενον ἄχρι τινὸς ἐν τινι πλέγματι διὰ τὸ θηριῶδες εἰρ⌐μένον. ἐπεὶ δὴ ἀπέθανε, περιχύσαντες αὐτὸ οἱ κομίζοντες, θεάμαl@ παρέχειν ἀπειθὲς εἰκόνα, μέχρι τῆς Κωνσαντίνε διεσπούσατο πόλεως. Καὶ μοι δοκεῖ τὸ ζῶν τὅτο Ἕλληνες πάλαι ἰδεῖν, ἢ ἐκπλαγέντες τῷ ξένω τῆς θέας, Θεὸν σφίσιν νομίσαι, εἰθισμένων αὐτοῖς τὰ παράδοξα θεοποιεῖν. Ὥσπερ ἢ τ̄ Σάτυρον. i. e. *This Ape-form is mixt with other* Species *of Animals; and this is plain, several being sent over to us; as that which is called* Pan, *which in its Head had a Goat's face and Goat's horns, from it's Loins downwards Goat's Legs; but in it's Belly, Breasts, and Hands was a pure Ape. Such an one the King of* India *sent to Constantius. This Animal lived for some time, and was carried about inclosed in a Cage, being very wild. When it died, those that looked after it, having embalmed it to make a shew of this unusual sight, sent it to* Constantinople. *Now I am apt to believe the ancient* Greeks *had seen this Animal, and being surprised at the strangeness of the sight, fancied it to be a God; it being usual for them to make a God of any thing that they admired or wondered at: as they did the* Satyr.

'Tis evident therefore by this Relation, that the *Satyr* is of the *Ape* or *Monkey-kind :* For πίθηκ@ here is generical, and includes both. But there being several *Species* of them, they received a denomination according to the resemblance they had to other *Animals*; as in *Philostorgius* are mentioned before, the *Leontopithecus*, the *Arctopithecus*, the *Cynocephalus* and *Aegopithecus*, which last seems to be our *Satyr*, from the resemblance it hath in it's Head and Legs to a *Goat.* That their Legs and Face are like a Goat's, is easie enough to be believed : but the *Horns* that they clap upon his Head, seem to me as an addition of the *Poets*, or the *Painters*, or both. But what gave a foundation to this Invention, possibly may be the large-

(l) *Bochart. Hierozoic. part. poster.* lib. 6. cap. 7. p. 829. (m) *Philostorg. Hist. Ecclesiastic.* lib. 3. cap. 11.

ness

nefs of their Ears ftanding off from their Head, and which are very remarkable. And this *Phurnutus* (*n*) gives as the reafon of it. *Horace* (*o*) takes notice of their Ears, but ill defcribes them in making them fharp pointed, whereas they are round.

————*& aures*
Capripedum Satyrorum acutas.

But by this Account it likewife appears, that *Pan* was a Name of this fort of *Monkey* ; and *Philoftorgius*'s Remark at the Conclufion of this Paffage, I think is very juft : for 'tis certain that this *Animal* was worfhipped in *India* as a *Deity*, as a Dog was by the *Ægyptians* ; and 'twas Death for any Body to kill one of them : For thus faith (*p*) *Diodorus Siculus* , Τὰς τε γδ αὐτὰς οἰκίας οἱ πίθηκοι καθάπερ τοῖς ἀνθρώποις, θεοὶ παρ' αὐτοῖς νομιζόμενοι, καθάπερ παρ' Αἰγυπίοις οἱ κύνες ἔκ τε τῶν παρεσκdυασμένων ἐν τοῖς τεμείοις τὰ ζῶα τὰς τροφὰς ἐλάμβανον ἀκωλύτως ὁπότε βέλοιτο. Καὶ τὰς προσηγορίας δι' ἐτίθεσαν γονεῖς τοῖς παισὶ κατὰ τὸ πλεῖστον ἀπὸ τῶν πιθήκων, ὥσπερ παρ' ἡμῖν ἀπὸ τῶν θεῶν. Τοῖς δι', ἀποκτείνασι τῦτο τὸ ζῶον, ὡς ἠσεβηκόσι τὰ μέγιστα, θανατθα ὥρισο πρόςιμεν. Διὸ δὴ καὶ παρὰ τιων εὐίχυσιν ἐν παροιμίας μέρει λεγόμενον ὅτι τῶν ἀναζευομένων, ὅτι πιθήκε αἷμα πολίσειαν. i.e. *Eafdem e-nim domus Simiæ quas Homines frequentant ; & pro Diis habentur apud illos, ut apud Ægyptios Canes ; paratos etiam in Cellis penariis cibos, quando libet, nemine prohibente, hæ beftiæ fumunt, nominaque ut plurimùm à Simiis, ut apud nos à Diis, Parentes Liberis fuis imponunt. Qui Animal hoc interfecerint, in eos, ut nefariæ Impietatis reos, fupplicio capitis animadvertitur. Ideo apud nonnullos Proverbij vicem obtinuit, quod in magnifice fe efferentes dicitur ; Simiæ Cruorem bibifti.* And in another place (*q*) *Diodorus* tells us, that *Pan* was in the greateft Veneration amongft the *Ægyptians*, and his Statue was in every *Temple*. And (*r*) *Juvenal* remarks,

Effigies Sacri nitet aurea Cercopitheci.

The Superftition of worfhipping this *Animal* obtained not only amongft the Ancients, but there are Inftances likewife of a later date, and what (*s*) *Johannes Linfchoten* relates, is very remarkable. ' How that in ' the Year 1554. the *Portuguefe* having taken the Ifland of *Ceylon*,they pro- ' pofed to rob a Temple on the top of *Adam*'s Pike ; but they found no- ' thing there,but a little Cabinet adorned with Gold and Jewels,in which ' was kept the *Tooth* of an *Ape*, which they took away, to the great grief ' of the Kings of that Place ; who fent Ambaffadors to the *Portuguefe*,and ' offered them Seventy thoufand Ducats for the *Tooth* ; which the *Por-* ' *tuguefe* were willing enough to take, but were diftwaded from it by ' their Bifhop *Gafpar*, who told them, that it was a Crime, thus to encou- ' rage the Idolatry of the *Indians* ; whereupon he burnt the *Tooth*, and ' flung the Afhes into the River. *Joh. Eufeb. Nierembergius* (*t*) hath the

(n) *Phurnutus de Nat. Deorum. Cap. de Pane.* p. m. 71.   (o) *Horace Odarum* , Lib. 2. Ode 19. (p) *Diodor.Sicul.Biblioth.Hift.*l.20.p.m.753.   (q) *Diodor.Sicul.ibid.*l.1.p.m. 16.   (r) *Juven. Satyr.* 15. v. 4. (s) *J.Linfchoten apud Theod.de Bry India Ori.*it.part.2.cap.45.p.m.111.   (t)*J.Euf.Nieremberg.Hift.Nat.*l.9. cap.45.p.180.
fame

fame Story, but varies in the Account of fome Particulars. And *Joh. Petrus Maffeius* (u) gives us a Relation of one of their Temples, which for Magnificence, might vie with any at *Rome:* His words are thefe; *Sanè fanum eſt Simiæ dicatum : Cujus duntaxat Pecori in victimarum uſum cuſtodiendo, Porticus miram in longitudinem excurrit, Columnarum Septingentarum è Marmore, tantæ magnitudinis, ut Agrippæ Columnas, quæ in celeberrimo quondam omnium Deorum Templo Romæ viſuntur, ſine dubio adæquent.* Now thefe Animals being worfhipped by the *Indians* as *Gods,* 'tis natural to believe (as *Lucretius* fuggefts) that they would invent and relate prodigious things concerning them ; and no doubt this gave the occafion to the *Poets* and *Hiſtorians* of making fuch fabulous Reprefentations of them: How far the latter might be concerned in the addition of *Horns* to the *Satyrs* Heads, I fhall not at prefent enquire : I call it an Addition, becaufe there is no Account from any credible Author, that there were ever obferved any of the *Ape-kind* to have Horns. Poffibly fome ancient Statues or Paintings might give fome light into this matter : for the ancient *Statuaries* and *Painters* were curious in reprefenting them ; and *Pliny* recommends, as excellent in this kind, the *Satyr* of *Mylo,* of *Lucippus, Antiphalus, Protogenes, Ariſton,* and *Nicomachus,* as Pieces admired in thofe days.

Albertus Magnus (w) who was happier in guefling, than in proving or defcribing what he meant ; tells us indeed, that the *Satyr* (whom he calls *Piloſus*) was of the *Ape-kind* ; but he makes fuch an odd Compofition of him, that one would take it to be rather a *Chimæra,* than a real Being: You may fee his words in the Citations.

(u) *Joh.Pet.Maffeij Hiſt.Indic.lib.1.p.m.36.* (w) *Albert.Magnus de Animalib. lib. 22. p. m. 223. Piloſus eſt Animal Compoſitum ex homine ſuperius, & Capra inferius ; ſed Cornua habet in fronte ; & eſt de genere Simiarum ; ſed multum monſtruoſum ; & aliquoties incedit erectum, & efficitur domitum. Hoc aſſerunt in Diſertu habitare Æthiopia ; & aliquoties captum & in Alexandriam deductum,& mortuum ſale infuſum & in Conſtantinopolin delatum.*

Of

Of the *SPHINGES* of the Ancients.

WE come now in the last place to discourse of the *Sphinges* of the *Ancients*, where I shall not relate all that is said of them; nor concern my self with the *Mythology* or Interpretation of the several *Fables* that have been invented about them; but I propose rather to shew, that there were indeed such *Animals* which the *Ancients* call'd *Sphinges*; and that they were not *Men*, but *Brutes*; and that they were of the *Ape* or *Monkey-kind*.

If we consult the fabulous Descriptions that are given of the *Sphinx*, we shall find it a very monstrous Composition. *Apollodorus* (a) tells us, ἐπεμψε γ᾽ Ἥρα Σφίγζα, ἣ μητρὸς μὲν Ἐχίδνης ἰῶ, Πατρὸς δὲ, Τυφῶνος, εἴχε δὲ πρόσωπον μὲν γυναικὸς ςῆθ. δὲ ἣ Βίαν κ᾽ ὀυρὰν λίον.,κ᾽ πίερυζας ὀρνιθ. *That* Sphinx *was the daughter of* Echidna *and* Typho, *she had the face of a Woman, the Breast, Feet, and Tail of a Lion, and the Wings of a Bird.* And (b) *Ausonius*,

> *Terruit Aoniam, Volucris, Leo, Virgo triformis*
> *Sphinx, volucris pennis, pedibus fera, fronte Puella.*

But as their Fansies govern'd them, so they made their description. *Clearchus* (as I find him quoted in (c) *Natalis Comes*) has out-done them all; *At* Clearchus (faith he) *Caput & Manus Puellæ, Corpus Canis , vocem Hominis, Caudam Draconis, Leonis ungues, Alas Avis, illam habuisse scripsit.* *Palæphatus* (d) is somewhat different in his Account, where he tells us, Περὶ τῆς Καδμείας Σφιγγὸς λέγεσιν ὡς Θηρίον ἐγένετο, σῶμα μὲν ἔχον ὡς κυνὸς, κεφαλὴν δὲ, κ᾽ πρόσωπον κόρης, πίερυζας ὀρνιθ., φωνὴ δὲ ἀνθρώπε· i. e. *They say that the* Cadmean Sphynx *was a wild Beast, having the Body of a Dog, the Head and Face of a Virgin, the Wings of a Bird, and the Voice of a Man.* But for the most part they make the *Sphinx Biformis* with a Maiden's Face and Lion's Feet; as the *Scholiast* upon (e) *Euripides* gives it, πρόσωπον παρθένε ςῆθ. κ᾽ πόδας λέοντος. So the *Scholiast* upon (f) *Aristophanes*, τί δ᾽ Σφιγγὸς πόδας λεοντάδεις ἦσαν· And *Euripides* himself, as he is quoted by (g) *Ælian*, makes her to have the Tail and Feet of a Lion, in that Verse,

> Ὀυρὰν δ᾽ ὑπειλᾶσ᾽ ὑπὸ λεοντόπεν βάσιν·
> *Caudam remulcens ad Leoninos pedes.*

Where we may observe that *Ælian* tells us here that the *Ægyptian Statuaries*, and the *Theban Fables*, made the *Sphinx* to be only *Biformis : Biformicm nobis conantur repraesentare, ipsam ex Corpore Virginis & Leonis cum gravitate compositam architectantes*, as Gesner there translates him: But the *Greeks* represented the *Sphinx* with wings; for as (h) *Ælian* in another place tells

(a) *Apollodori Bibliothec.*l.3.c.5.§8.p.m.170.   (b) *Ausonius in Grypho Ternarij.*   (c) *Natalis Comes Mytholog.*l.9.c.18.   (d) *Palæphatus de incredibilibus Historiis Cap.de Sphinge.*p.m.14.   (e) *Eurip. in Phanissis.* (f) *Aristophan. in Ranis.*   (g) *Ælian.de Animal.*l.12.c.7.   (h) *Ælian.de Animal.*l.12.cap.38.

us,

us, *Sphingem quicunque vel Pictura vel Plastica opcram dant, fingere alatam solent.*

But our chief Bufinefs is to enquire, how *Nature* hath formed them ; and not how the *Poets* , *Painters* , or *Statuaries* have, according to the Luxuriancy of their Fancie, feigned or figured them ; to. fhew what they really are in themfelves, and not what *Hieroglyphically* the Ancients might intend or underftand by them; and we fhall find, that they are only a fort of *Ape* or *Monkey*, that is bred in *Æthiopia* and amongft the *Troglodytes*, of a comely Face, with long Breafts, thence up to their Neck not fo hairy as on the reft of their Body ; and are of a mild and gentle Nature. For thus (i) *Pliny, Lynces vulgo frequentes, & Sphinges, fufco Pilo, mammis in Pectore geminis Æthiopia generat.* And fo (k) *Solinus, Inter Simias habentur & Sphinges,villofæ comis* (*Salmafius* reads it *villofæ omnes*) *mammis prominulis ac profundis, dociles ad feritatis oblivionem. Ælian* (l) places them amongft the wild Beafts of *India*, where he tells us , *Naturali quodam Ingenio & Prudentia valent etiam apud nos Animalia , non totidem tamen, quot funt in India : illic enim hujufmodi funt, Elephantus, Pfittacus, Sphinges & nuncupati Satyri, & Indica Formica.* And *Artemidorus* in (m) *Strabo* tells us, that the *Sphinges, Cynocephali* and *Cepi* are bred amongft the *Troglodytes.* *Agatharchides* (n) confirms the fame,and gives us this Account of them ; Αἱ σφίγες ἐ, οἱ Κωνοκάφαλοι ἐ, Κῆροι πας̃απέμπον) εἰς τω̃ 'Αλεξανδρείαν ὲκ τ̃ Τρωγλοδυτικῆς, ἐ, τ̃ Αἰθιοπίας· εἰση) αἱ μὲν Σφίγες ταῖς γεαφομέναις παρόμοιαι. Πλω ὃτι πᾶσαι δασεῖαι, ἐ, ταῖς ψυχαῖς ἥμεροι ἐ, πεα̃οι. Καὶ πανεργίας κοινωνῦσι πλείσης, διδασκαλίας τε μεθοδ̃ λικῆς ὲπι ποσὸν ἁπτον), ὥσε τω̃ εὐρυθμίαν ὲν πᾶσι θαυμάζειν. i. e. *The Sphinges, Cynocephali and Cepi are fent to Alexandria from the Country of the Troglodytes and* Æthiopia. *The Sphinges are like to what they are painted,only they are all hairy, and mild and gentle in their Nature : they have a great deal of Cunning, and a Method of Learning what they attain to, that one would wonder at their aptnefs to any thing.* *Diodorus Siculus* (o) gives us much the fame Relation, and 'tis likely *Agatharchides* borrowed his from him : for he tells us, Αἱ ὃ Σφίγες γίγνον) μὲν πεεί τε τω̃ Τρωγλοδυτικω̃, ἐ, τω̃ Αἰθιοπίαν, ταῖς ὃ μορφαῖς ὠπάρχωσιν ὲκ ανόμοιοι ταῖς γεαφομέναις, μόνον ὃ ταῖς δασύτησι διαλλάτⁱσι. Τὰς ὃ ψυχὰς ἡμέρας ἔχεται ἐ, πανέργες, ὲπι πλεῖον ἐ, διδασκαλίαν μεθοδ̃ικω̃ ὲπιδέχον). i. e. *Sphinges circa Troglodyticam & Æthiopiam exiftunt, formâ his non abfimiles, quæ Arte Pictorum exhibentur, nifi quôd hirfutia tantummodò differunt. Placidi illis funt Animi, & verfuti, artifque quæ compendio tradi folet, admodum capaces.* But *Philoftorgius* (p) is fo particular in his Defcription, and he is the more to be credited, becaufe he declares he had feen them himfelf, that I think I need

---

(i) *Plinij Hift. Nat.* lib.8.cap.21.p.m.168. (k) *Solinus Polyhift.* cap.27. p.m.39. (l) *Ælian. de Animal.* lib.16.cap.15. (m) *Strabo Geograph.* lib.16.p.533. (n) *Agatharchides apud Photij Biblioth.* p. m. 1362. cap. 38. (o) *Diodorus Siculus Biblioth.* lib.3. p. m.167. (p) *Philoftorgius Hift.Ecclefiaft.* l.3.c.11. p.41.

no more Authorities to prove what I have here laid down, that these
Sphinges were only a sort of *Ape* or *Monkey.* I will therefore give you
his own words, which are these ; Καὶ μὲν ἡ Σφὶγξ γένΘ. ἔςι πιθήκων (αὐτὸς
γ᾽ ΔεσπάμενΘ. γράψω) ἧς τὸ μὲν ἄλλο σῶμα λάσιόν ἔςιν, ὡς τοῖς ἄλλοις πι-
θήκοις· τὸ δ᾽ ςέρνον ἄχρι γε αὐτῆ τῆ τραχήλε ἐψίλωJ, μᾳῖ᾽ς δ᾽ γυναικὸς ἔχει.
Ἐρυθρὰ τινὸς βραχίΘ. καὶ χεροειδῆς ἐπανατήμαJΘ. ἅπαν εἰς κύκλω τὸ γεγυμνω-
μένον τῷ σώματι τ᾿ Θ. περιδέων᾽Θ., καὶ εἰς πολλ᾽ύ τινα εὐπρέπειαν ἀνθρωποφανῆ
ὄντι τῷ ἐν μέσω χρώματι συναρπαζομένη. Τὸ δ᾽ πρόσωπον ἀνερρογύλω J μᾶλλον, κ᾿ εἰς γυναικίαν ἕλκει μορφήν. Ἡ τὸ φωνὴ ὑποτεικὰς ἀνθρωπεία, πλὴν ὅσον
ἐκ εἰς ἄρθρα διαιρεμένη, ἀλλά τινι παχέως. καὶ ὀ᾽ς μετά τινΘ. ὀργῆς τὸ κ᾿ ἀγθη-
δόνΘ. ἀσημα ᾿αποφθεγγομένης προσεικότα βαρύτερα τε μᾶλλον ἔςιν ὀξυνομένα·
ἀγέλον τὸ ἔςι δεινῶς τὸ θηρίον, κ᾿ πανηργότατον, κ᾿ οὐδὲ ῥαδίως τιθασευόμενον.

i. e. *A Sphinx is a sort of Ape (I shall write what I saw my self) all the rest
of whose Body is hairy like other Apes. But it's Sternum or Breast is smooth
without hair up to the Throat. It has Mammæ or Breasts like a Woman ;
little reddish Pimples like Millet Seeds, running round that part of the Body
that is bare ; very prettily suiting with the Flesh colour in the middle. It's
Face is roundish, and resembles a Woman's. It's Voice is very much like the
Humane, only it is not articulate, but præcipitate ; and like one that speaks
unintelligibly thro' Anger and Indignation. When 'tis incensed, it's Voice is
deeper. This Animal is very wild, and crafty, and not easily tamed.* And
Pierius, as I find him quoted by (q) *Philip Camerarius,* gives us much the
same description of one he saw at *Verona. Harum ego unam* (saith he)
*Veronæ quum essem vidi ; Mammæ illi & Glabris & Candidis , à Pectore
propendentibus. Circumducebat eam circulator quidam Gallus, ex ignotis
antea Insulis recens advectam.* And a little after adds , *Ipsa verò Sphinx
toto erat pectore glabello, facie & auribus humanis proprioribus , dorso hispido
supra modum, fusco & oblongo Pilo, eoque densissimo.*

What has been said, I think fully makes out, that the *Sphinx* is not a
meer Figment of the *Poets,* but an *Animal* bred in *Africa,* of the *Ape* or
Monkey-kind. 'Tis different from our *Orang-Outang* in the colour of it's
Hair ; in the roundness and comeliness of it's Face ; in it's Breasts, being
pendulous and long ; and the red Pimples it hath on the naked part of
it's Body. *Pliny* tells us (as I have elsewhere remarked) that the *Sphinges* have Pouches in their Chops as *Satyrs* and *Monkeys* have ; and the
Poets describing them with a Lion's Tail , make me apt to think , that
they are of the *Monkey-kind.*

(q) *Phil. Camerarij Opera subcisiva sive Meditat. Hist. Cent.* 1. *Cap.* 71. p. m. 325.

F I N I S.

An Advertisement

Of some Discourses and Observations made by Dr. *Edw. Tyson*, and where published.

PHOCÆNA, or the *Anatomy* of a *Porpess*, dissected at *Gresham-Colledge* ; with a *Preliminary* Discourse concerning *Anatomy* ; and a *Natural History* of *Animals, Lond.*Printed for *Benj. Tooke* at the Ship in St. *Paul's* Church-yard, 1680. in 4ᵗᵒ.

Vipera Caudisona Americana ; or the *Anatomy* of a *Rattle Snake* ; dissected at the Repository of the *Royal Society*, *Jan.* 168¼. vide *Philosoph. Transactions* Nᵒ 144. p. 25.

Lumbricus Latus, or a Discourse read before the *Royal Society* of the *Joynted Worm.* Wherein a great many Mistakes of former Writers concerning it, are remarked : it's Natural History from more Exact Observations is attempted : and the whole urged, as a Difficulty, against the Doctrine of Univocal Generation. Vide *Philosoph. Transactions* Nᵒ 146. pag. 146.

Lumbricus Teres,or some Anatomical Observations on the *Round Worm*, bred in Humane Bodies. Vide *Philosoph. Transactions* Nᵒ 147. pag. 154.

Tajaçu, sive Aper Mexicanus Moschiferus ; or the *Anatomy* of the *Mexico Musk-Hog*. Vide *Philosoph. Transact.* Nᵒ 153. pag. 359.

Lumbricus Hydropicus, or an Essay to prove, that *Hydatides* often met with in Morbid Bodies, are a *Species* of *Worms*, or *imperfect Animals*. Vide *Philosoph. Transact.* Nᵒ 193. pag. 506.

Carigueya, seu Marsupiale Americanum ; or the *Anatomy* of an *Opossum*, dissected at *Gresham-Colledge.* Vide *Philosoph.Transact.* Nᵒ 239. pag.105.

Ephemeri Vita, or the Natural History and Anatomy of the *Ephemeron* ; a Fly that lives but five hours. Written originally in Low-Dutch; by *Jo. Swammerdam* M. D. of *Amsterdam*, and published in English by *E. Tyson* M. D. *Lond.* Printed for *Henry Faithorne* and *John Kersey* at the Rose in St. *Paul's* Church-yard. 1681. in 4ᵗᵒ.

Embrionis Galei lævis Anatome. Vide *Franc. Willoughbæi Hist. Piscium. Edit. à Jo. Raio in Appendic.* pag. 13.

Lumpi Anglorum Anatome. ibid. pag. 25.

The *Scent-Bags* in *Poll-Cats*, and several other *Animals*,first discovered. Vide Dr. *Plot's* Natural History of *Oxfordshire*, pag. 305.

Vide *Thom. Bartholini Acta Medica & Philosophica Hafniensia, Vol.* 5. ubi,

Observ. 26. *Vomica Pulmonis.*

1

Observ.

Obſerv. 27. *Hydrops Thoracis, & difficultatis ſpirandi rara Cauſa.*

Obſerv. 28. *Hæmoptoe, Tuſſis, Pleuritis & Empyema à duobus claviculis, fortuitò in Pulmones delapſis.*

Obſerv. 29. *Polypus omnes Corporis totius Venas & Arterias occupans.*

Obſerv. 30. *Polypus Bronchiarum & Tracheæ.*

Vide ejuſd. Obſerv. 101. *Obſerv.* 107. *Obſerv.* 108.

Some Anatomical Obſervations of *Hair* found in ſeveral Parts of the Body ; as alſo *Teeth, Bones,* &c. with Parallel Hiſtories of the ſame obſerved by others. *Vide* Dr. *Hooks Philoſophical Collections* N° 2. pag. 11.

Anatomical Obſervations of an *Abſceſs* in the *Liver* ; a great number of Stones, in the *Gall-bag* and *Bilious Veſſels* ; an unuſual Conformation of the *Emulgents* and *Pelvis.* A ſtrange *Conjunction* of both *Kidneys,* and great Dilatation of the *Vena Cava. Vide Philoſoph. Tranſact.* N° 142. p.1035.

An Anatomical Obſervation of four *Ureters* in an Infant ; and ſome Remarks on the *Glandulæ Renales.* ibid. pag. 1039.

An Abſtract of two Letters from Mr. *Sampſon Birch* an Alderman and Apothecary in *Stafford,* concerning an Extraordinary Birth ; with Reflections thereon. *Vide Philoſoph. Tranſact.* N° 150. pag. 281. and Dr. *Plot's Natural Hiſtory* of *Staffordſhire,* pag. 272.

The Figure of the *Cochineal* Fly. *Vide Philoſoph. Tranſact.* N° 176. pag. 1202.

An Obſervation of *Hydatides* found in the *Veſica Urinaria* of Mr. *Smith. Vide Philoſoph. Tranſact.* N° 187. pag. 332.

An Obſervation of an Infant, where the *Brain* was depreſſed into the Hollow of the *Vertebræ* of the Neck. *Vide Philoſoph. Tranſact.* N° 228. pag. 533.

An Obſervation of one *Hemiſphere* of the *Brain* ſphacelated ; and of a *Stone* found in the Subſtance of the *Brain. Vide Philoſoph. Tranſact.* N° 228. pag. 535.

www.ingramcontent.com/pod-product-compliance
Lightning Source LLC
Chambersburg PA
CBHW021705210326
41599CB00013B/1533